Laboratories, workshops, and sites.

Concepts and practices of research in industrial Europe, 1800–1914

Robert Fox

and

Anna Guagnini

Office for History of Science and Technology
University of California, Berkeley
1999

COVER ILLUSTRATION: The laying of the transatlantic telegraph cables was conducted under the constant supervision of electrical engineers, who performed a great deal of testing work on board cable-laying ships. The illustration of "Awaiting the reply," an oil painting by Robert Charles Dudley, shows the testing cabin of the *Great Eastern*. Reproduced with permission of the Metropolitan Museum of Art, gift of Cyrus W. Field, 1892 (92.10.43).

Berkeley Papers in History of Science, 18
Uppsala Studies in History of Science, 27
Bologna Studies in History of Science, 6
Historical Studies in the Physical and Biological Sciences
29:1 (1998) and *29:2* (1999)

Order from
Office for History of Science and Technology
University of California
543 Stephens Hall # 2350
Berkeley, CA 94720-2350
Price $24.00

ISBN: 0-9672617-0-8
Library of Congress catalog card number: 99-63056

Laboratories, workshops, and sites.

Concepts and practices of research in industrial Europe, 1800–1914

Robert Fox
and
Anna Guagnini

In memory of

Donald Cardwell

1919–1998

CONTENTS

LIST OF ILLUSTRATIONS

PREFACE

Laboratories, workshops, and sites is the product of a program of comparative research on which, among our various other commitments, we have been engaged for the last ten years. In that time, we have incurred many debts, most notably for the financial support we have received successively from the Leverhulme Trust, the Research and Equipment Committee of the University of Oxford, and the Centro Internazionale per la Storia delle Università e della Scienza (CIS) of the University of Bologna. We have also benefited from our association with the Achievement Project, an international study of cultural innovation financed by the Renaissance Trust (1991–5), and from grants and other material assistance that we have received from the Royal Society and the Deutsches Museum, Munich, as well as from the libraries and archives in which we have worked.

The responses of colleagues, too, have been important, and we are glad to express our gratitude to the participants in seminars in Oxford, Warwick, Milan, Barcelona, Eindhoven, and Baltimore and at the Dibner Institute in the Massachusetts Institute of Technology for the generosity and frankness of their comments. Among the reactions we have particularly valued, however, special mention must be made of those we received over a long period from Graeme Gooday and other colleagues and students in Oxford. Another important source of encouragement and criticism was the Second International Summer School in History of Science in Uppsala in July 1990. This book, in fact, is based on an extensive reworking of the lectures we gave on that occasion. We are grateful to the organizers of the Summer School, in particular to our main host in Uppsala, Töre Frangsmyr, for the opportunity of attempting a first ordering of our material. It is a particular pleasure, therefore, to see *Laboratories, workshops, and sites* appearing in the series of publications through which the Office for History of Science and Technology of the University of California, Berkeley, along with its partners in Uppsala and Bologna, has diffused the work of the summer schools—now held at two-year intervals in Berkeley, Bologna, and Paris as well as Uppsala—beyond those able to attend.

Finally, we extend our thanks to those who have helped during the preparation of the present text for publication. From the start, Kluwer and the Banca di Roma as copyright-holders generously gave us free rein in the use of material (now revised and incorporated in parts of chapters 3 and 4) that we published in a very provisional form in Peter Kroes and Martijn Bakker, eds., *Technological development and science in the industrial age. New studies on the science-technology relationship* [Boston studies in the philosophy of science, vol. 144] (1992), and *The journal of European economic history*, 23 (1994), 69–92. At a later stage, John Heilbron offered encouragement and characteristically searching criticism of a draft of the complete work and arranged for the preliminary publication of the text in two parts in *Historical studies in the physical and biological sciences*, 29, part 1 (1998) and 29, part 2 (1999).

As our thoughts took shape, Stephanie Jenkins in the Modern History Faculty in Oxford prepared successive versions of the book with unfailing accuracy and cheerfulness. The Faculty also helped with a grant that made it possible for Dr Katherine Watson to assist us with some of the last editorial tasks, most notably with the preparation of the index.

Oxford and Bologna

Robert Fox and Anna Guagnini
October 1999

INTRODUCTION

One of the attractions of applied research as a field for historical inquiry is that it lies at the interface of several disciplines, drawing equally on the history of science and technology, economic and business history, social history, and science policy. Perhaps for this reason, it has been difficult to achieve a broad common perspective that goes beyond individual case studies with specific disciplinary orientations. Nevertheless, because of its intrinsic interest and its policy implications, the subject has attracted sustained attention for almost half a century. In this period, the volume of literature has grown rapidly. One particularly favored quarry has been the synthetic dyestuffs and pharmaceutical industries in the late nineteenth and early twentieth centuries: as a result, we know a great deal about the research strategies of Bayer, BASF, Hoechst, and the other leading German manufacturers and about their contribution to the spectacular expansion that David Landes, as long ago as 1965, termed "Imperial Germany's greatest industrial achievement."[1] Other industrial sectors, other periods, and other countries, too, have attracted attention, though mainly with regard to the period since the first world war, when (most notably in the U.S.) in-house laboratories and the mechanisms for systematic consultation with academic scientists became a routine, if by no means universal, feature of company structure, contributing to the model of the "prototypes of the modern industrial enterprise" that Alfred Chandler has analyzed in his influential writings.[2]

A related but quite distinct focus has been the history of laboratories within the academic world and of their evolving links with industrial pro-

1. David S. Landes, *The unbound Prometheus. Technological change and industrial development in Western Europe from 1750 to the present* (Cambridge, 1969), 276; the phrase we cite first appeared in the earlier version of this work, published in *The Cambridge economic history of Europe, 6:1* (Cambridge, 1965), 503.
2. See, for example, Alfred D. Chandler, Jr., *Scale and scope. The dynamics of industrial capitalism* (Cambridge, MA and London, 1990), passim; the phrase we cite appears on p. 11. Chandler finds his "prototypes" in the United States, Great Britain, and Germany since the first world war.

duction. Here too chemistry has attracted a great deal of attention, much of it directed to the tradition of laboratory-based research that took root in the German universities following the establishment of Justus von Liebig's laboratory at the University of Giessen in 1824.[3] More recently, the history of physics laboratories has also become a focus for scholarly activity, yielding a wealth of insights on the functioning and multiple roles of such laboratories since they began to proliferate in universities and other institutions of higher education in Europe and North America in the 1850s and 1860s. This historical work, with its strong emphasis on experimental practices and instruments, is especially relevant to the present book for the light it has thrown on the contribution of academic physics laboratories to such rapidly growing industrial specialties as telegraphy and optics. Most importantly, it has shown that the bonds between the academic world and manufacturing were closer than earlier writings have suggested. In *Laboratories, workshops, and sites,* we build on these newer departures in the secondary literature and on the lines of interpretation they offer.[4] In doing so, we pay special attention to the variety of motives and circumstances that favored the interaction between science and industry, in particular though by no means exclusively with reference to the electrical industry. We devote a good deal of space, for example, to our argument that the growth of physics laboratories in the later nineteenth century owed much to their perceived value for industry and to the utilitarian air that, at least in this

3. On the practices of Liebig's laboratory and their significance, see Jack B. Morrell, "The chemist breeders: the research schools of Liebig and Thomson," *Ambix, 19* (1972), 1–46; Frederic Lawrence Holmes, "The complementarity of teaching and research in Liebig's laboratory," in Kathryn M. Olesko, ed., *Science in Germany. The intersection of institutional and intellectual issues,* published as *Osiris,* 2nd ser., *5* (1989), 121–164; and William H. Brock, *Justus von Liebig. The chemical gatekeeper* (Cambridge, 1997), esp. chapts 2, 3.

4. The literature in this area, to much of which we refer in the chapters that follow, is extensive and growing. Among the books of special relevance to our point are: Robert H. Kargon's pioneering, *Science in Victorian Manchester. Enterprise and expertise* (Manchester, 1977); Crosbie W. Smith and M. Norton Wise, *Energy and empire. A biographical study of Lord Kelvin* (Cambridge, 1989); and David Cahan, *An institute for an empire. The Physikalisch-Technische Reichsanstalt 1871-1918* (Cambridge, 1989). Also highly relevant are two articles by Graeme Gooday: "Precision measurement and the genesis of physics teaching laboratories in Victorian Britain," *The British journal for the history of science, 23* (1990), 25-51, and "Teaching telegraphy and electrotechnics in the physics laboratory: William Ayrton and the creation of an academic space for electrical engineering in Britain 1873-1884," *History of technology, 13* (1991), 73-112. See too the several discussions of the commercial and scientific impact of the development of sophisticated quantitative methods, especially for the determination of physical constants, in M. Norton Wise, ed., *The values of precision* (Princeton, NJ, 1994).

period (though less obviously after the turn of the century), pervaded most of them. We also explore what we see as the still rather neglected subject of the ways in which the organization of laboratories and research was approached in schools and departments of engineering.[5] Here, however, we should be the first to recognize that despite the richness of the relevant recent studies, generalization remains difficult; our own contribution, therefore, has a provisional character that reflects the amount of work that remains to be done.

While our starting point in what we say about chemistry as well as about physics and the relations between these sciences and industry lies in interpretations emerging from the secondary literature, we try to extend the implications of existing historical work in what we see as some of its most fruitful directions. One such direction lies in the emphasis that has been placed on the routine character of much laboratory activity, such as the staple procedures of analysis and quality control in chemistry and of precise measurement, the reproduction and control of standards, and calibration in physics. Another, related direction is the one suggested by Thomas Hughes's observation that "the most active research site" in the American electrical industry at the end of the nineteenth century was the manufacturing company, whose engineering departments performed the role later assumed by research-and-development laboratories.[6] Our own conception of research, like Hughes's and (with reference to telegraphy) Paul Israel's,[7] embraces under that term a wide range of activities extending far

---

A contribution of great interest is Simon Schaffer's "Late Victorian metrology and instrumentation: a manufactory of Ohms," in Robert Bud and Susan E. Cozzens, eds., *Invisible connections. Instruments, institutions, and science* (Bellingham, WA, 1992), 23-56. Schaffer draws attention to the "practical" rhetoric that pervaded the experimental work carried out in the Cavendish Laboratory and elsewhere.

5. Two especially important recent studies that touch on this subject are Wolfgang König, *Technikwissenschaften. Die Entstehung der Elektrotechnik aus Industrie und Wissenschaft zwischen 1880 und 1914* (Chur, 1995), and König, "Science-based industry or industry-based science? Electrical engineering in Germany before World War I," *Technology and culture*, 37 (1996), 70–101.

6. Thomas P. Hughes, *Networks of power. Electrification in western society, 1880–1930* (Baltimore, MD and London, 1983), 160. While we accept Hughes's point, we should want to stress, with reference to Europe, that in many industries, including the electrical industry, research and development performed in workshops and on the site remained important well after 1900.

7. Paul Israel, *From machine shop to industrial laboratory. Telegraphy and the changing context of American invention, 1830–1920* (Baltimore, MD and London, 1992), 185–186, where the author criticizes the tendency of historians to identify the work of industrial research with laboratories "based primarily on scientific, academic models."

beyond the narrow definition of research as original work performed in a laboratory and geared to and producing novelty. For us, in fact—and here our position is that of most of the actors we treat—prominence has to be given not only to laboratory-based research (whether research leading to fundamental new departures or the more routine work of precise physical measurement) but also and no less centrally to research conducted on the site (such as Sebastian Ziani de Ferranti's daring trial runs at his Deptford power station after its construction in 1889) and to the often obscure and always laborious processes by which known technologies were adapted to particular purposes and contexts.

The various categories of research to which we refer were by no means water-tight. The permeability of the boundaries between the categories is illustrated, for example, by the way in which essentially routine work led, on occasions, to marketable innovations: the (admittedly rather few) synthetic dyes that were developed commercially from the Bayer company's extensive base in the day-to-day work of patenting and quality control are a case in point.[8] In electricity, the improvements in the generation and transmission of alternating current that Charles E.L. Brown introduced during the 1890s in his capacity as a technical director of the Swiss Maschinenfabrik Oerlikon (and later at Brown, Boveri et Compagnie) are an equally striking illustration of the capacity of a long train of minor, very practical improvements, many of them made on the site, to yield substantial advances.[9]

The fluidity of the forms that applied research, in the sense of an experimental activity explicitly oriented towards practical objectives, took was especially evident before the first world war, on which we have chosen to focus this book. Indeed, it was precisely the unformed, multifaceted character of the notion of applied research and of the contexts in which it was pursued that drew us to the subject. In trying to break with the preoccupation of most of the existing literature with laboratory-based research as the mainspring of a change in industrial technology, we also sought to address the recurring broader question of the place of science in the early history of what is, commonly but often confusingly, referred to as science-based industry. Wolfgang König's recent reformulation of the question in terms that ask us to embrace the concept of an "industry-based science" and the notion of a mutual (as opposed to a uni-directional) interaction between science and electrical engineering in the late nineteenth century is a helpful reminder of the fragility of the very category of science-based industry[10]—a category that has taken a severe historiographical buffeting ever

8. See pp. 263-266.
9. Hughes, *Networks of power* (ref. 6), 131–139.
10. König, *Technikwissenschaften* (ref. 5), 227–296, and "Science-based industry or industry-based science?" (ref. 5).

since the journal *Technology and culture* emerged as a platform for a series of contributions (by Derek De Solla Price, Edwin T. Layton, and others) questioning the dependence of technology on science and seeking to highlight the distinctive character of technology as a form of knowledge.[11]

Another characteristic of the secondary literature that has weighed with us is the growing interest, especially among economic and business historians, in countries other than those, such as Germany and the U.S., that are conventionally placed in the "fast lane" of industrial development.[12] In extending our discussion to attitudes towards research in countries that in our period found themselves in the "slow lane"—notably Britain, France, and Italy—we are not attempting to establish a league-table of success and failure, still less to advance an ideal model or universal prescription for success. Our aim is rather to identify the particular forms that applied research took in geographical, political, and economic contexts far removed from those of Germany and the U.S. It would not have been realistic for a country as industrially backward as Italy, for example, to try to compete at the most advanced frontiers of technology. The problem for Italy was not simply the adequacy of academic laboratories or of the training available to engineers—the factors to which contemporaries most commonly ascribed the success of Germany and the U.S. It lay rather in the small size of the Italian industrial base, which made the accumulation of technical know-how difficult and an investment in innovative research impossible to justify. Nevertheless, as we argue, these circumstances did not prevent a significant involvement in a process of adaptation which, although it had little

11. See especially Derek J. De Solla Price, "Is technology historically independent of science? A study in statistical historiography," *Technology and culture, 6* (1965), 553–568; Edwin T. Layton, "Mirror-image twins: the communities of science and technology in 19th century America," *Technology and culture, 12* (1971), 562–580, and "Technology as knowledge," *Technology and culture, 15* (1974), 31–41. The discussion has been continued by Walter Vincenti in his *What engineers know and how they know it. Analytical studies from aeronautical history* (Baltimore, MD, 1990). Other important contributions to the debate, cast with special reference to the United States, are George Wise, "Science and technology," in Sally Gregory Kohlstedt and Margaret W. Rossiter, eds., *Historical writing on American science,* published as *Osiris,* 2nd ser., *1* (1985), 229–246; John M. Staudenmaier, S.J., *Technology's storytellers. Reweaving the human fabric* (Cambridge, MA and London, 1985); and Rachel Laudan, "Natural alliance or forced marriage? Changing relations between the histories of science and technology," *Technology and culture, 36* (1995), 17–28.

12. For our distinction between countries in the "fast" and "slow" lanes, see Robert Fox and Anna Guagnini, "Life in the slow lane: research and electrical engineering in Britain, France, and Italy, c.1900," in Peter Kroes and Martijn Bakker, eds., *Technological development and science in the industrial age. New studies on the science-technology relationship* [Boston studies in the philosophy of science, volume 144] (Dordrecht, Boston, and London, 1992), 133–153.

in common with institutionalized laboratory-based research, must, in our analysis, still be regarded as a form of research.

In a book of necessarily limited size, we have had to make choices about the industrial sectors that we treat. Here, as in any study of the history of applied research in the nineteenth and early twentieth centuries, organic chemistry had to be given prominence as the area of manufacturing in which the debate about the way in which science should be marshaled in the structures of industry initially took shape: our first chapter, in which early developments in dyestuffs, especially after the introduction of the first synthetic dye, mauve, in 1856, figure prominently, is informed by this consideration. In the later chapters, we do not lose sight of chemistry, but the main emphasis shifts to electricity and the growth of the electrical construction and supply industry between the 1880s and 1914. While in this part of the book we have been able to draw on some excellent studies, we are conscious of the tendency of historians to give pride of place to developments in the U.S. and to finely cast case-histories which, though often penetrating, do not lend themselves to generalizations going beyond the specific American experience. For this reason, we have deliberately directed our attention primarily to Europe and adopted an approach that associates occasional detailed consideration of specific examples with an attempt to identify broader, transnational trends.

In conclusion, while we fully acknowledge our debt to existing writing, both as a resource in its own right and as a guide to our own archival research, we return to our conviction about the unevenness of what is available to us. We have been especially conscious of the gaps in one little-explored area, to which we refer above: that of research performed within industry but in engineering workshops and on sites that lay beyond the walls of the conventional laboratory. Here records, as well as the secondary literature, strike us as conspicuously thin, for both fast-lane and slow-lane countries. We know little, for example, of the pioneering work on high-voltage alternating current generators performed at AEG, Ganz, or Brown, Boveri, and biographies of those who conducted such work are few and far between. Our book is not written as an answer or remedy for this deficiency, but it is written in full consciousness of it. The gap is, of course, far from being the only major one. But we are convinced that until it is addressed, the historical debate on the relations between science and the development of science-based industry (however we choose to construe that term) is likely to be seriously impaired. We may have moved definitively from over-simple models, based on the passage of scientific innovations from the realm of theory and experiment, via a reformulation as "applied science," to that of industrial practice. But we have yet to exploit fully the much richer historiographical visions now open to us, which demonstrate (in a manner exemplified in Walter Vincenti's *What engineers know and how they know it* and Ulrich Wengenroth's *Enterprise and technology*[13]) the importance of innovative work carried on the shop floor and underline the need for such work, which overlapped with and often coexisted with that of what we now identify as in-house research laboratories, to be examined in the depth it deserves.

13. Vincenti, *What engineers know* (ref. 11), and Ulrich Wengenroth, *Enterprise and technology. The German and British steel industries, 1865–1895,* trans. [from the original German edition of 1986] by Sarah Hanbury Tenison (Cambridge, 1994).

1 CONCEPTIONS OF APPLIED RESEARCH BEFORE 1870

By the second half of the nineteenth century, there existed throughout the industrial world an almost universal consensus that the practices of industry would benefit from a more intimate association with science. The belief, and its corollary that institutions should be created or adapted to facilitate the association, was not a novel one. It had roots going far back to medieval discussions of the relations between *scientia* and *ars*, the realms of knowing and doing. And thereafter, during the Renaissance, it had been given practical expression in the convergence of the activities of scholars and craftsmen—a convergence reflected in the Baconian aphorisms asserting the oneness of knowledge and power. But it was during the eighteenth century that the belief had begun to gain wide currency. Larry Stewart's *The rise of public science* and *Public & private science* by Alan Morton and Jane Wess have explored the form that the belief took in England in the early years of the century: as these works show, and as the refurbished George III Collection at the Science Museum in London illustrates spectacularly, lecturing, scientific publishing, and instrument-making for demonstration and other purposes were all stimulated in this period by a sense of the value of science, in particular of Newtonian science, for the emerging industrial economy.[1] Hence by the time Denis Diderot argued, in the great French *Encyclopédie* in 1751, that the arts (in the broad sense that embraced technology) could only progress through a union of "practice" and "speculation," his was already by no means a lone voice.[2] But over the next half-century, the dream of a fertile marriage of learned and practical cultures that he articulated recurred with increasing frequency in writing on industrial improvement and began to make a decisive mark on the realities of both science and manufacturing.

1. Larry S. Stewart, *The rise of public science. Rhetoric, technology and natural philosophy in Newtonian Britain* (Cambridge, 1992), and Alan Q. Morton and Jane Wess, *Public & private science. The King George III Collection* (Oxford, 1993).
2. Denis Diderot, "Art," in *Encyclopédie, ou dictionnaire raisonné des sciences, des arts et des métiers* (Paris etc., 17 vols. plus plates, supplements, and tables, 1751–1780), *1* (1751), 714.

This vision of the potential value of knowledge made its mark in countries across Europe as monarchs and governments implemented policies aimed at advancing their economic and strategic interests.[3] It is not surprising, however, that the vision took root most easily in France. There, the soil had long been fertilized by a tradition of governmental involvement in manufacturing going back to the later seventeenth century, to the time of Louis XIV's minister Jean-Baptiste Colbert. The state enterprises of the Gobelins (for tapestries and carpets), Saint-Gobain (for glass), and Sèvres (for porcelain) all routinely engaged *savants* as well as artisans in the elaboration of new techniques. They drew on and helped to foster a "bookish" interest in technology which, in the reigns of Louis XV and Louis XVI (1715-1793), produced a succession of masterpieces of technological description extending chronologically from René Antoine Ferchault de Réaumur's *L'art de convertir le fer forgé en acier* (1722) to the plates and related articles on technical subjects in the *Encyclopédie* (1751-1780) and the many volumes of the magnificent compendium of trade and handicraft practices, the *Description des arts et métiers*, that the Académie des Sciences began publishing, after several decades of delay, in 1761.[4]

The growth of interest in the relations between the work of the *savant* and that of the *artiste* during the Ancien Régime is a natural starting point for a study of the emergence of the concepts and practices of what eventually crystallized as industrial research. The road that led from Diderot's programmatic statements to the structures within which science came to be integrated in manufacturing in the late nineteenth and early twentieth centuries was to be a tortuous one: it was littered with the difficulties that contemporaries experienced in determining the mechanisms by which theoretical and experimental science might exert a significant influence on the procedures of the factory or workshop, as opposed to simply describing or analyzing them. In a manner typical of the beginning of the process, the French chemist Pierre-Joseph Macquer saw his function in writing on the dyeing of silk in the 1760s as essentially that of a diffuser of the craft procedures he had witnessed in the workshop of one of the leading silk dyers of

3. On this process and its very uneven character, see Robert Fox, "Science and government in eighteenth-century Europe," in Roy S. Porter, ed., *The Cambridge history of science*. Vol. 4. *Science in the eighteenth century* (Cambridge, in press).
4. On the content and publishing history of the *Descriptions*, see Arthur H. Cole and George B. Watts, *The handicrafts of France as recorded in the* Descriptions des arts et métiers *1761–1788* [Kress Library of Business and Economics, Publication no. 8] (Boston, 1952). On the utilitarian interests of the Académie des Sciences, see Roger Hahn, *The anatomy of a scientific institution. The Paris Academy of Sciences, 1661–1803* (Berkeley, 1971), passim in chapters 1–5, and Robin Briggs, "The Académie Royale des Sciences and the pursuit of utility," *Past & present*, no. 131 (May 1991), 38–88.

the day—procedures that would normally have been kept hidden from the public or even from other dyers.[5] But gradually, over the next century and a half, the limited descriptive function that Macquer conceived for himself gave way to far stronger claims: first, that science could provide a means of controlling processes and making them more efficient, and then that it could yield not just economies or minor improvements in existing practice but also novelties from which totally new methods, new products, and even new industries could be developed. The early stages of that transition, which was to bring science from the margins to the center of modern industrial society, provide the central theme of this chapter.

Defining roles: Manufacturers and consultants

Diderot's ideal began most obviously to be realized in a concrete form in the last two decades of the Ancien Régime, in ways consistent with the centralized, hierarchical structures of pre-revolutionary France.[6] Overwhelmingly, this occurred at the higher levels of power and influence. The main agents were not employees (who tended to remain loyal to older, traditional practices whose evolution was barely touched by science) but rather those who, in varying ways, were in a position to control industrial activities. Such figures would typically be representatives of the technical arm of the state, independent entrepreneurs, or *savants* whose authority rested on their demonstrated scientific eminence. At least in France, therefore, science in this period entered the industrial sphere through men who, as established "insiders" in the worlds of administration, industry, or learning, had no need to justify their activity, in the way that the new and very different breed of academically trained industrial scientists and engineers in the mid and later nineteenth century had to do.

Macquer and other chemists, such as Claude-Louis Berthollet, Antoine Lavoisier, and Jean-Antoine Chaptal, were typical of eminent *savants* who combined their work as teachers and men of science with engagements as regular part-time advisers on practical matters, mainly though not exclusively to the state manufactories.[7] A more unusual but by no means unique case was that of Jean Hellot. A man of literary and other talents who trained informally as a chemist, Hellot went on to become a productive member of

5. Pierre-Joseph Macquer, *Art de la teinture en soie* (Paris, 1763), iii–iv.
6. For an excellent account of the relations between science, industry, and the French state in these decades, see Charles Coulston Gillispie, *Science and polity at the end of the Old Regime* (Princeton, NJ, 1980), esp. (on the examples discussed in this and the following paragraph) 335–548.
7. On this facet of the work of Berthollet, Lavoisier, and Chaptal, see Michelle Sadoun-Goupil, "Science pure et science appliquée dans l'oeuvre de Claude-Louis Berthollet," *Revue d'histoire des sciences*, 27 (1974), 127–145; Patrice Bret, "Lavoisier

the Académie Royale des Sciences. But he was also, and simultaneously a notable authority on a variety of metallurgical and other technological practices who accepted virtually full-time employment, first as the government's inspector general of dyeing (in which capacity he published a notable work on the theoretical and practical aspects of the dyeing of wool) and then, from 1751, as technical adviser to the porcelain factory at Vincennes (later at Sèvres, when the factory moved on receiving the title "royal").[8] In a manner particularly characteristic of France, engineers from the various state *corps* also contributed to the growing recognition of the relevance of their abstract knowledge to practice. In this, the education offered at the schools that fed the *corps* played a crucial role. From its formation in 1747, for example, the Ecole Royale des Ponts et Chaussées provided both incomparable training and a setting for important work by its professors on the strength of materials and other aspects of structural engineering.[9] It was here that Gabriel Jars was able to refine the experience he had obtained as an observer of his father's mining enterprises at Sainbel and Chessy near Lyon: the benefits of the two years he spent at the school (although he did not enter the *corps*) were to be seen in a classic work, *Voyages métallurgiques*, in which he described and analyzed the practices he had witnessed on governmental missions across northern and central Europe in the 1750s and 1760s.[10] By the eve of the French Revolution, the recently founded Ecole Royale des Mines had also emerged as a significant contributor, with its pupils and teachers continuing the efforts of Jars and others to bring French metallurgical and mining techniques up to the standards of those being developed in Britain and Germany.

Any scepticism about the industrial value of science that remained by the 1790s was removed, at least in French eyes, by the mobilization of *savants* and engineers in the revolutionary and Napoleonic wars that set

---

et l'apport de la chimie académique à l'industrie des poudres et salpêtres," *Archives internationales d'histoire des sciences*, 46 (1996), 57–83; and John Graham Smith, *The origins and early development of the heavy chemical industry in France* (Oxford, 1979), passim, but esp. chapter 2.

8. "Eloge de M. Hellot," in *Histoire de l'Académie Royale des Sciences. Année 1766* (Paris, 1769), 167–179. Hellot's treatise appeared as *L'art de la teinture des laines et des étoffes de laine en grand et petit teint, avec une instruction sur les débouillis* (Paris, 1750).

9. Antoine Picon, *L'invention de l'ingénieur moderne. L'Ecole des Ponts et Chaussées 1747–1851* (Paris, 1992), esp. 53–207.

10. Gabriel Jars, *Voyages métallurgiques ou recherches et observations sur les mines et forges de fer...faites en 1758, jusques & compris 1769, en Allemagne, en Suede, Angleterre, Norvege, Tirol, Liege, & en Hollande* (3 vols., Lyon and Paris, 1774–81). For a biographical sketch of Jars, see Granjean de Fouchy's "Eloge de M. Jars" in vol. 1 of the *Voyages métallurgiques*, xxi–xxviii.

France against Britain and her allies almost without a break from 1792 to 1815.[11] It was now generally accepted as a matter of strategic as well as commercial import that manufacturing should become more "scientific" in its approach. Under the pressure of war and a maritime blockade that deprived French manufacturers of many of their normal raw materials, the rhetorical value of such technological triumphs as the production of artificial soda and sugar beet and the devising of new methods for preparing gunpowder (in the absence of Chilean nitrates) was exploited to the full. Sometimes the claims for what the Bureau de Consultation des Arts et Métiers and other committees of the government and the Institut de France had achieved verged on the extravagant. But there was much justice in the sense of pride that suffused Chaptal's retrospective account of the contributions of science to French industry during the long years of war. For Chaptal, writing in 1819, it was the progress of chemistry, allied to accelerating mechanization, that had made the French response to the rigors of the Continental Blockade possible.[12] In the mid-eighteenth century, as he argued, French manufacturers had rightly rejected the advances of the *savant*.[13] But, following the chemical revolution that Lavoisier and his circle had brought to pass in the 1770s and 1780s, chemistry had advanced at last to the status of a "positive science," thereby opening the way for an integration in the world of practice that had already borne rich fruit.[14]

Despite the many questions that were begged by Chaptal's evaluation of the contribution of the new chemistry to the economy, by 1819 he could look back on a period of nearly thirty years in which, spurred on by the extraordinary circumstances of war and the Blockade, the technological roles of manufacturers and scientific consultants had become significantly more prominent, in private as well as in state-controlled industry, than they had been even at the end of the Ancien Régime. His own career, which took him from an education in chemistry and medicine and a base in the academic environment of Montpellier to an ever greater engagement in manufacturing, illustrated one of the main channels by which a *savant* could engage with industry.[15] By the time he began investing his private fortune in the manufacture of mineral acids, soda, and other chemicals in the Midi

11. In a rich literature on the mobilization, see for example: Joseph Fayet, *La Révolution française et la science 1789–1795* (Paris, 1960), chapter 3, and Nicole and Jean Dhombres, *Naissance d'un nouveau pouvoir. Sciences et savants en France (1793–1824)* (Paris, 1989), esp. chapter 10.

12. Jean-Antoine Chaptal, *De l'industrie françoise* (2 vols., Paris, 1819), *2*, 1–113.

13. Ibid., *2*, 36–50.

14. Ibid., *2*, 38.

15. On Chaptal's career and industrial activity, see Smith, *Origins and early development of the heavy chemical industry in France* (ref. 7), esp. 22–24, 59–61, and 77–78.

in the mid-1780s, he was already well established in the chair of chemistry that he held in Montpellier, under the auspices of the regional assembly (the *Etats de Languedoc*), and he retained his foot-hold in the academic world long after his definitive removal to Paris in 1797. There, he continued to straddle the worlds of the *savant* and the *artiste*. His subsequent position as a member of the distinguished scientific circle over which Berthollet and the mathematician Pierre-Simon Laplace presided at Arcueil during the First Empire[16] was entirely compatible with his authorship of a four-volume work on applied chemistry addressed to those working in the chemical industry.[17]

Of the other manufacturers of the period who built industrial careers in France on a background and continuing interest in scientific investigation, none enjoyed Chaptal's prominence or intellectual distinction. But about the turn of the century and on into the early nineteenth century, it became increasingly common for some kind of training in science, in particular in chemistry or the scientific aspects of medicine, to be seen as a natural prelude to entrepreneurial activity. A former *polytechnicien* and assistant to Guyton de Morveau, Charles Bernard Desormes, and Nicolas Clément, the first holder of the chair of industrial chemistry at the Conservatoire Royal des Arts et Métiers from 1819, were characteristic of this expanding generation of scientifically informed men who combined the skills and outlook of the chemist and the entrepreneur; for many years they worked together in the manufacture of alum at Vesoul.[18] Equally typical was the somewhat younger Henri Schlumberger, who attended courses in chemistry in his native Mulhouse and Paris (albeit briefly), after joining the printing firm of Nicolas Koechlin et frères as an apprentice colorist at the age of fifteen (in 1818).[19]

Although France was undoubtedly the pace-maker, it was not only there that the role of industrialists and employees who had had a significant, though often informal, exposure to science became more prominent during the first half of the nineteenth century. In England, as in all the industrialized countries of Europe, a similar change took place. There, however, the integration of the leading industrial science, chemistry, in manufacturing—chiefly in the dye-factories and color shops of the textile areas—proceeded rather slowly. The remarkable career of the colorist John Mercer

16. On the Arcueil circle, see Maurice P. Crosland, *The Society of Arcueil. A view of French science at the time of Napoleon I* (London, 1967).
17. Jean-Antoine Chaptal, *Chimie appliquée aux arts* (4 vols., Paris, 1807).
18. Charles Dunoyer, "Nécrologie. Clément-Desormes," *Journal des économistes, 1* (1841–1842), 208–213.
19. A. Scheurer-Rott, "Notice nécrologique sur M. Henri Schlumberger," *Bulletin de la Société Industrielle de Mulhouse, 24* (1852), 115–127, on 116–117.

indicates how haltingly the process occurred in Lancashire. When Mercer entered the textile industry in his late teens in 1809, he took the traditional route of a very practical apprenticeship in the color shop of the Oakenshaw print-works in Lancashire.[20] Nine years later, when he moved up to the newly created position of "experimental chemist" in the same color shop (after a period in which he reverted to his original trade as a hand-loom weaver), only two other firms in the whole of the British calico-printing industry employed anyone who possessed, in his words, "a knowledge of chemistry."[21] In fact, Mercer was self-taught as a chemist: his knowledge of the subject was refined informally through a reading of textbooks and contacts with other chemists, in particular, in his case, with a group that met each month in the early 1840s at Whalley, near Clitheroe, and through his day-to-day industrial work.[22] By the 1850s, however, when Mercer retired as a national figure honored by his election as a Fellow of the Royal Society (in 1852) but still unmistakably an autodidact, the small handful of scientifically competent experts in the textile industry, with backgrounds and experience very different from Mercer's, had begun to grow.

Robert Bud and Gerrylynn Roberts have analyzed these developments against the background of an emerging consensus on the boundaries that separated pure from applied science and on the benefits of an easy interaction between the two realms.[23] As they have shown, even in England, where scepticism with regard to science had always been more marked than in France, a broad swath of industry around the mid-century became more receptive to chemists who had passed through higher education than it had been during Mercer's early career or *a fortiori* that of his older contemporary, Samuel Parkes, the chemical manufacturer and author of the widely read *Chemical catechism* (1806) and *Chemical essays* (1815).[24] Among these new recruits was a small but by no means negligible number trained in the

20. On Mercer's career, see Edward Andrew Parnell, *The life and labours of John Mercer, F.R.S., F.C.S. etc. The self-taught chemical philosopher* (London, 1886). Mercer's work as a colorist is well explored in Agustí Nieto-Galan, "Calico printing and chemical knowledge in Lancashire in the early nineteenth century: the life and 'colours' of John Mercer," *Annals of science*, 54 (1997), 1–27.
21. Parnell, *Life of Mercer* (ref. 20), 241.
22. Ibid., 103–104. See also Robert H. Kargon, *Science in Victorian Manchester. Enterprise and expertise* (Manchester, 1977), 89–90. The "Whalley meetings" took place in the house of Lyon Playfair, who was working in Clitheroe at the time, and at his suggestion.
23. Robert Bud and Gerrylynn K. Roberts, *Science versus practice. Chemistry in Victorian Britain* (Manchester, 1984), esp. chapters 2–4.
24. Frederick Kurzer, "Samuel Parkes: chemist, author, reformer—a biography," *Annals of science*, 54 (1997), 431–462. Although self-taught like Mercer, Parkes was

German chemical laboratories that were founded during the second quarter of the century. Lyon Playfair, who was employed by the Clitheroe calico-printer James Thomson in 1841, fresh from taking a Ph.D. under Justus von Liebig at the University of Giessen, is the best-known of the new generation, although he spent barely a year with Thomson before moving on to the position of honorary professor of chemistry at the Royal Institution in Manchester and in due course to the post of chemist to the Geological Survey and a brilliant career in academic and public life in London.[25] Similar but more enduring mid-century commitments to industry were those of two other products of the Giessen laboratory, Thomas Richardson and Edmund Ronalds. Richardson studied with Thomas Thomson in Glasgow before taking a doctorate at Giessen and working briefly in the important private laboratory of the chemist Théophile-Jules Pelouze, a friend and former pupil of Liebig, in Paris.[26] But on his return to his native Newcastle-on-Tyne, he devoted himself entirely to the applications of chemistry and to chemical manufacturing, in particular the purification and production of soft lead and the manufacture of superphosphates by a method proposed by Liebig. Despite his appointment first at the Newcastle-upon-Tyne School of Medicine and Surgery in 1848 and then, from 1856, as reader in chemistry at the University of Durham, Richardson's industrial interests remained undimmed, and he continued to work and publish on applied subjects, in particular on the qualities of different types of steam coal, until his death in 1867. Ronalds, by contrast, pursued an academic career for sixteen years before turning to manufacturing. After complementing his studies at Giessen with periods in other German laboratories and in Zurich and Paris, he held teaching posts, between 1840 and 1856, first at St Mary's Hospital and the Middlesex Hospital in London and then as professor of chemistry at Queen's College, Galway, and he was already in his mid-thirties by the time he took over a chemical works near Edinburgh for the processing of the waste tar products and ammonia liquor left from the manufacture of gas.[27] He worked there for almost twenty years, maintain-

---

convinced of the value of chemical instruction for those engaged in the chemical industry at all levels; see Parkes, *An essay on the utility of chemistry to the arts and manufactures* (London, 1807).

25. Wemyss Reid, *Memoirs and correspondence of Lyon Playfair, First Lord Playfair of St. Andrews P.C., G.C.B., LL.D., F.R.S., &c.* (London, 1899), esp. chapters 3, 4, and 5 on Playfair's education, life in Lancashire, and early career in London.

26. Unsigned obituary notice of Thomas Richardson in Royal Society of Edinburgh, *Proceedings, 6* (1866–1869), 198–9.

27. J.Y. Buchanan, "Dr Edmund Ronalds," Royal Society of Edinburgh, *Proceedings, 17* (1889–1890), obituary notices, xxviii–xxix; also the brief notice in *Journal of the Chemical Society. Transactions, 57* (1890), 456.

ing his chemical interests and eventually, in retirement, devoting himself entirely to research in his private laboratory.

As Mercer observed with reference to his own domain of dyeing and dye-making, the trend that Playfair, Richardson, and Ronalds exemplified was sufficiently strong in Britain to produce signs of resentment, on the part of the older dyers and colorists, against the new generation of those whom they disparagingly described as "the chemical men."[28] In many cases, the taunts must have rung true. An academic education without a thorough immersion in the practical problems of the trade was simply not enough to allow the newcomers to compete on equal terms with the traditional aristocracy of practitioners, whose knowledge and skills had been fashioned amid the complex, if rough and ready practices of natural dye chemistry. What was required was limited and specific: it came down to economy and, above all, consistency in the composition of the dyes and reliability in the colors, with innovation being restricted to relatively minor variants on existing recipes and practices. For this purpose, experience of the kind that Mercer acquired in abundance during his long career, rather than sophisticated theory, was the main requirement.

The character of the evolving relations between the academic and the shop-floor traditions in the mid-nineteenth century varied so markedly with the country, the sector, and the readiness of individual industrialists to adjust their procedures that simple generalization is far less useful than a more flexible form of analysis of the kind explored by F.L. Holmes in his discussion of a multiplicity of distinct "cultures of chemistry."[29] However, the tensions that accompanied the changes we have described were universal. Throughout industry, in fact, the growing prominence of men with a scientific background was viewed with a mixture of defensiveness and suspicion by the skilled workmen who had relied on experience and the rules of thumb that were now increasingly under scrutiny. A good example, especially in Britain, was that of the iron and steel industry. Here, a reticence similar to that of the older generation of colormen was encapsulated in the attitude that Sir Henry De la Beche criticized in his inaugural discourse at the opening of the School of Mines and of Science applied to the Arts (later to become the Royal School of Mines) in London in November 1851. According to De la Beche, who had been since the 1830s a leading pioneer of instruction and research in geology (as founder and director of

28. Parnell, *Life of Mercer* (ref. 20), 245.

29. Frederic Lawrence Holmes, "Beyond the boundaries: concluding remarks on the workshop," in Bernadette Bensaude-Vincent and Ferdinando Abbri, eds., *Lavoisier in European context. Negotiating a new language for chemistry* (Canton, MA, 1995), 267–278, on 274–277. For a comment on this approach, see Nieto-Galan, "Calico printing and chemical knowledge," (ref. 20), 2.

both the Museum of Practical—later Economic—Geology and the Geological Survey), scepticism with regard to the relevance of science to practice was rife in Britain. There were even those who thought a familiarity with science a disadvantage:[30]

> We still...too frequently hear of practical knowledge as if in a certain sense opposed to a scientific method of accounting for it, and as if experience, without that advantage, were more trustworthy than the like experience with it.

Soon afterwards, John Percy took up the theme in his introductory lecture for the course in metallurgy at the same institution.[31] Like most advocates of a more scientific approach to technology, he insisted on the necessity of experience of full-scale production, without which in metallurgical establishments "most erroneous and occasionally ruinous results" could be obtained.[32] But he was equally adamant that only scientific knowledge could temper the theories of the practical man, which he described as "often in the highest degree wild and visionary."[33] It was this kind of knowledge that his course would provide, through lectures and laboratory instruction in assaying and analytical chemistry that would help to achieve the combination of the scientific with the practical on which improvements in metallurgy depended.

The assertiveness of Percy's response reflects the provocation offered to someone with a solid background in science (obtained, in his case, through a full training as a doctor at Edinburgh) by those whose defence of empiricism was more a matter of pride than of embarrassment. And Percy was not alone. By the mid-century, men of science everywhere were ready to insist, in public, on their usefulness in the industrial world, and frequently (in the manner of Macquer, Berthollet, and Chaptal) they moved from precept to practice by assuming the role of consultant or independent entrepreneur at the same time as they held some of the most elevated positions in the scientific community. It was while he occupied the chair of chemistry at the Sorbonne, for example, that Joseph-Louis Gay-Lussac accumulated a number of paid appointments as an industrial advisor. The income from these appointments became important to him during the Bourbon

30. Henry T. De la Beche, "Inaugural discourse, delivered at the opening of the School of Mines and of Science applied to the Arts, 6th November 1851," *Records of the School of Mines and of Science applied to the Arts, 1:1* (1852), 1–48, on 20.
31. John Percy, "On the importance of special knowledge to the practical metallurgist," ibid., 127–148.
32. Ibid., 134.
33. Ibid., 136.

Restoration and even more so under the July Monarchy: in the 1830s and 1840s, he acted as a regular consultant to the Saint-Gobain glassworks, earning fees of almost £300 a year, and even more lucratively as Assay Master at the Mint, a post for which he received about four times as much.[34] In addition, there were occasional exceptional earnings, most notably between 1844 and 1846, when he received £7000 for the sale of the rights to his Gay-Lussac tower for preventing the escape of the noxious oxides of nitrogen that were produced in the manufacture of sulphuric acid.[35] Clearly, for him as for many other academic scientists, such as Louis-Jacques Thenard, Victor Regnault, and Théophile-Jules Pelouze in France, Michael Faraday and Charles Wheatstone in Britain, and Justus von Liebig in Germany, there was no reticence about an engagement in industrial or (in the case of Liebig or the French chemist Jean-Baptiste Boussingault[36]) agricultural practice. Work of this kind, like that generated at about the same time for analytical chemists by the battle against the scandals of urban squalor, environmental pollution, and the adulteration of food,[37] was seen as part and parcel of a career in science in the industrial age.

Academic conceptions

From a very early stage, the enthusiasm with which academic scientists came to direct their knowledge and energy to industrial problems was routinely qualified by a clear statement of the terms on which most of them conceived their engagement. While always ready to parade their familiarity with the realities of factory life, they presented themselves above all as the theorists of industry, who would go beyond the superficial descriptions of the traditional manuals and books of secrets to elucidate the principles underlying practice, but do so without becoming wholly assimilated in the industrial world. In the early and mid-nineteenth century, chemists made the point most frequently and consistently. As Chaptal observed in 1807, the capacity of chemistry to aid industry would be undermined if the distinctive roles of the chemist and the manufacturer became confused:

34. Maurice P. Crosland, *Gay-Lussac. Scientist and bourgeois* (Cambridge, 1978), 178–204 and 228–234.

35. Ibid., 203.

36. Liebig's contributions to agricultural chemistry are well covered in William H. Brock, *Justus von Liebig. The chemical gatekeeper* (Cambridge, 1997), 145–182. On Boussingault's work and life, see Frederick W.J. McCosh, *Boussingault. Chemist and agriculturalist* (Dordrecht, Boston, and Lancaster, 1984).

37. Colin A. Russell, with Noel G. Coley and Gerrylynn K. Roberts, *Chemists by profession. The origins and rise of the Royal Institute of Chemistry* (Milton Keynes, 1977), 103–109.

"the chemist and the manufacturer can be of mutual assistance to each other, but it is essential that each of them should retain the place that the character of their respective fields of study has designated for them."[38] A similar point was made two decades later by the rising young chemist Jean-Baptiste Dumas, who used the preface of his eight-volume *Traité de chimie appliquée aux arts* (1828) to justify the amount of "pure" chemistry that he had included in the work, despite its title.[39] For Dumas, the effective application of chemistry in industry demanded a thorough familiarity with the basic principles of the subject: theory, as he saw it, should therefore be regarded as an indispensable element, one that a younger generation of manufacturers and industrial employees in particular could not afford to ignore.

The perception of August Wilhelm Hofmann, one of Liebig's ablest and most successful pupils, was similar. Hofmann nailed his colours to the mast as someone whose roots remained firmly in academic chemistry, in his case at the Royal College of Chemistry in London, where he had been director and professor of chemistry since the foundation of the college in 1845. His position as one of the senior jurors at the International Exhibition of 1862 in London lent authority to his view that the lures of utility and material profit had become so intrusive that they risked undermining the distinctive quality of the chemist's contribution. After a review of the scientifically based successes of the exhibition in the field of artificial dyestuffs, which he proffered as "a splendid demonstration of the power and utility of chemistry," Hofmann referred to "another and a nobler lesson" that was to be learned from Faraday's wholly disinterested discovery of benzol (later revealed to be an essential ingredient of aniline dyes) in 1825. The lesson, he wrote,[40]

> is apt to be forgotten in a great industrial country like England where fortune ever tempts the chemist from the hard mountain path of pure scientific research to the smiling slopes of lucrative industrial application. That lesson, based on the discovery of benzol, and pointing to the illustrious name of its discoverer FARADAY, bids us, if we would reach the noblest heights of fame to seek PURE TRUTH; careless of industrial advantage to ourselves, yet sure that from our labours practical good will in due season flow, for the benefit of all mankind.

38. Chaptal, *Chimie appliquée aux arts* (ref. 17) *1*, xxi.
39. Jean-Baptiste Dumas, *Traité de chimie appliquée aux arts* (8 vols., Paris, 1828–1846), *1*, viii–ix.
40. A.W. Hofmann, report on Class II ("Chemical and pharmaceutical products and processes"), Section A ("Chemical products and processes"), in *International Exhibition, 1862. Reports of the juries on the subjects in the thirty-six classes into which the Exhibition was divided* (London, 1863), 136.

Hofmann's call was typical of the conditions that the more eminent scientists tended to impose in return for their collaboration.[41] As he saw it, the main weakness besetting the interaction between science and industry in the 1860s lay not in the detachment that writers since Diderot had deplored but rather in an undue proximity, amounting sometimes (at least on the part of chemists) to a servile infatuation. The most obvious safeguard against a too hasty union lay for him, as it did for Chaptal and Dumas before him, in the preservation of the scientific autonomy and theoretical orientation of the consultant's work. However, attempts to articulate precisely the way in which the maintenance of a resolute allegiance to the goal of "pure truth" would benefit industry were rare. It was one thing for Chaptal to create a vague link between the grandiose assertion that the late eighteenth and early nineteenth centuries had been a period in which "la science et la pratique se sont éclairées réciproquement"[42] and the industrial progress that he advanced as the direct consequence; it remained, and remains for historians today, quite another thing to unravel his view of the logical bond underlying the link. Clearly, the practice of chlorine bleaching would have been inconceivable but for the initial discovery of chlorine in Carl Wilhelm Scheele's laboratory and its development as a bleaching agent, through a sequence of systematic tests, by Berthollet. But it is far less evident that the discovery of the element and, *a fortiori*, the techniques for using it in bleaching owed anything to contemporary chemical theory.[43] Just how far Chaptal himself would have wished to press a claim for the value of theory (as opposed to a familiarity with chemical substances and a skill in systematic experimenting) in the technology of chlorine bleaching, at a time when the processes involved were equally obscure in both the old and the new chemistry, is unclear.[44]

41. Henry Roscoe made an equally clear statement of the primacy of "pure science" in his evidence to the Devonshire Commission in 1870, quoted in Bud and Roberts, *Science versus practice* (ref. 23), 85. His very active, life-long involvement in the bridging of science and industry, however, lends a certain rhetorical hollowness to his assertion.
42. Chaptal, *De l'industrie françoise* (ref. 12), 2, 39.
43. The distant bearing of contemporary theory on new departures in the eighteenth-century chemical industry is elaborated in two classic articles by Charles Coulston Gillispie: "The discovery of the Leblanc process," *Isis, 48* (1957), 152–170, and "The natural history of industry," ibid., 398–407. See also Gillispie, *Science and polity in France at the end of the Old Regime* (ref. 6), 411–412.
44. Chaptal's discussion of chlorine bleaching in his *Chimie appliquée aux arts* (ref. 17), 4, 194–197, is conspicuously brief. For modern studies that bring out the marginal role of theory and the centrality of scientifically informed craft practices, see, in addition to the articles by Gillispie cited in note 43, Barbara Whitney Keyser, "Between science and craft: the case of Berthollet and dyeing," *Annals of science, 47* (1990), 213–260, and Robert Fox, "Science, practice, and innovation in the age of

The fact remains, however, that, by the mid-nineteenth century and despite the imprecision of the supporting analysis, academic chemists had arrived at a strikingly cosy consensus on their value to industry. The consensus was so strong that when Lyon Playfair delivered a grave warning about the consequences of British industry's indifference to chemistry in the aftermath of the Great Exhibition of 1851, he was mouthing home truths that would already have been perfectly familiar to many in his progressive metropolitan audience at the Society of Arts, Manufactures, and Commerce.[45] In a manner that would have been endorsed not only by the more scientifically minded members of the society but also by the advocates of the industrial importance of chemistry from Chaptal to Hofmann, Playfair insisted that "abstract" rather than "practical" science constituted the "life and soul" of industry; hence it was the "cultivators of abstract Science, the searchers after truth, for eternal truth's own sake" whom it was important to support.[46] The view was one to which Playfair's mentor, Liebig, would also have warmed, even though Liebig's early initiatives in the training of chemists in analytical techniques since the 1820s had been primarily directed at the preparation of pharmacists (with a lesser but growing proportion of aspirants to academic careers) rather than of employees for manufacturing.[47]

Playfair used his rather woolly rhetoric to argue for a two-fold contribution that science, in particular chemistry, would make once it was properly harnessed. First and most importantly, as he maintained, chemists within

natural dyes, 1750–1860," in Maxine Berg and Kristine Bruland, eds., *Technological revolutions in Europe. Historical perspectives* (Cheltenham, 1998), 86–95.

45. Lyon Playfair, "The chemical principles involved in the manufactures of the Exhibition as indicating the necessity of industrial instruction," in *Lectures on the results of the Great Exhibition of 1851, delivered before the Society of Arts, Manufacturers, and Commerce*, 2nd ser. (London, 1852), 157–208. For an earlier statement of the importance of improving chemical education if Britain was to break with her tradition of "inefficient knowledge or manipulation" and avoid the snares of wastefulness and complacency with regard to her rivals in France and Germany, see the preface (by the editors, Edmund Ronalds and Thomas Richardson) to the English edition of Friedrich Knapp, *Lehrbuch der chemischen Technologie, zum Unterricht und Selbstudium*, published as *Chemical technology; or, chemistry applied to the arts and to manufacturers* (3 vols., London, 1848–1851), *1*, vi–viii.

46. Playfair, "The chemical principles" (ref. 45), 190–191.

47. J.B. Morrell, "The chemist breeders: the research schools of Liebig and Thomas Thomson," *Ambix*, *19* (1972), 1–46, on 17–26. As Morrell observes (ibid., 21–23), the number of chemists going on to industry from the laboratory of Thomas Thomson at the University of Glasgow (in the two decades or so following his appointment as professor of chemistry in 1818) seems to have been significantly higher.

or as advisors to industry would enhance profits and efficiency by controlling and recycling waste and testing the quality of products and ingredients. In gas lighting, for example, it was only the chemist who could identify and help to remove the impurities that gave rise to blackening and fumes;[48] and proof of the value of precise analysis already existed in experiments that showed that, until the intervention of chemistry, about a third of the coloring ingredient of madder was lost unwittingly and unnecessarily during the washing of dyed fabric.[49] The second, very different contribution was to the devising of totally new processes. Here, Playfair may in retrospect appear a visionary with an unusually clear-eyed perception of the new departures that were to transform the chemical industry over the next quarter of a century. But in the circumstances that prevailed when he was writing, he was on much shakier ground. His strategy was to move from a handful of examples intended to create the impression of the fruitfulness of chemistry—such as chlorine bleaching and Mercerization[50]—to a generalized assertion about the capacity of science to stimulate or accelerate innovations. As Playfair's evidence showed, many of these innovations had their roots in chance observations. But the absence of any hard and fast prescription that would help to multiply such observations did nothing to shake his confidence. Science remained essential, in his view, as the only body of knowledge that would allow the fruits of chance to be recognized and exploited to the full.

Despite his occasional more extravagant flourishes, Playfair's comments reflected the core reality that the industrial role of science in the early 1850s was still seen primarily as a matter of understanding leading to quality control, economy, and minor improvements, much as it had been since the later eighteenth century. Elsewhere in his address, however, Playfair articulated a less conventional vision when he broached the place that science should have in modern education. On this point, his views reflected trends that had begun to make their mark on industry in the decade or so before he spoke. They rested on the premise that industrial practice would be improved and innovation made easier by engaging a workforce that understood processes as opposed to merely performing them.[51] The premise had its banal side: it can be seen as an elaboration of the familiar rhetoric of the pioneers of lower and middle-level technical education who had established the Mechanics' Institutes in Britain, the *écoles d'arts et métiers* in France, and the *Gewerbeschulen* in Germany over the previous half cen-

48. Playfair, "The chemical principles" (ref. 45) 186–189.
49. Ibid., 173–174.
50. Ibid., 169–170 and 172.
51. Ibid., 193–207.

tury.[52] But while Playfair recognized the continuing importance of the education that these intellectually modest institutions provided, his prescription extended beyond them to a new generation of schools in which science would have the prestige of the classics and the other traditionally noble disciplines. Specifically, he advocated the creation of industrial universities and colleges of a standing comparable with that of his favored model, the Ecole Centrale des Arts et Manufactures in Paris, an advanced school in which senior engineers had been trained for industry since 1829.[53] From such a system, Playfair believed, manufacturers would be able to draw employees possessing an unprecedentedly sophisticated background in the industrially relevant sciences and a social status appropriate to a new learned profession.[54]

Playfair's confidence in the benefits that would flow from an expansion of the kind of instruction he prescribed was reinforced by developments in higher education which, despite their recent origin, were already irreversible and ubiquitous. Throughout Europe, in the 1840s and 1850s, a convergence of rapid industrial growth and the heightened aspirations of a nascent academic profession helped to make the mounting of new scientific courses with what was declared to be economic relevance a common priority of most educational reform movements. Even in Britain, where the establishment of a common cause capable of rallying not only the scientific community but also government and broad sections of the nation's cultural elite was particularly difficult, the signs of change were obvious:[55] here, conspicuous developments included the establishment of the Royal College of Chemistry in 1845 and the creation, in 1851, of the Government School of Mines and of Science applied to the Arts.[56] Likewise in France, in

52. On these earlier initiatives, see Donald S.L. Cardwell, *The organisation of science in England* (2nd edn., London, 1972), 39–44 and 71–75; Ian Inkster, ed., *The steam intellect societies. Essays on culture, education and industry circa 1820–1914* (Nottingham, 1985); Charles R. Day, *Education for the industrial world. The Ecoles d'Arts et Métiers and the rise of French industrial engineering* (Cambridge, MA, 1987); Peter Lundgreen, *Techniker in Preussen während der frühen Industrialisierung* (Berlin, 1975), 41–84.
53. Playfair, "The chemical principles" (ref. 45), 202. The standard general history of the Ecole Centrale remains Léon Guillet, *Cent ans de la vie de l'Ecole Centrale des Arts et Manufactures 1829–1929* (Paris, 1929). For a more recent study of the first fifteen years of the institution, see John Hubbel Weiss, *The making of technological man. The social origins of French engineering education* (Cambridge, MA, 1982).
54. Playfair, "The chemical principles" (ref. 45), 204.
55. On the mid-nineteenth-century origins of the concern about the state of science and the implications for British industry, see Peter Alter, *The reluctant patron. Science and the state in Britain 1850–1920* (Oxford, 1987), esp. 75–137.
56. See above, pp. 79-80, and (on the School of Mines) Margaret Reeks, *Register of the Associates and old students of the Royal School of Mines and history of the Royal School*

1846, Dumas, in his capacity as Dean of the Faculty of Science at the Sorbonne, advanced ambitious proposals for the reorganization of the teaching of the nation's eleven science faculties along lines that would take account of industrial and agricultural need rather than of traditional academic criteria, while remaining distinct from purely technical and shop-based instruction: there was to be a new *licence* and a new doctorate in "sciences mécaniques" and a compulsory practical examination for all candidates for the *licence*, except those specializing in mathematics.[57] Nothing was done at the time, and the revolutionary disturbances of 1848 further delayed change. But eventually, in the 1850s, the stirrings that Dumas represented bore fruit, albeit of a kind somewhat different from what he had proposed in 1846. The creation of five new faculties of science (at Clermont-Ferrand, Lille, Marseille, Nancy, and Poitiers) and a number of *écoles préparatoires à l'enseignement des sciences et des lettres* (essentially providing the first two years of the courses offered in the faculties of science and letters) significantly extended the opportunities available in the existing faculties and such technical schools as the Ecole Centrale des Arts et Manufactures and the four less advanced *écoles d'arts et métiers*) for those wishing to prepare for industrial careers.

In Germany and Switzerland, the new departures of the mid-century also made their mark. Here, for some decades political and economic circumstances had engendered a belief in the special efficacy of education as a motor of material advancement, and both technical schools (especially after the founding of the pioneering Polytechnische Schule in Karlsruhe in 1825[58]) and universities had prospered through a reform of all sectors of the educational system. As an abundant secondary literature has shown, chemistry was one of the main beneficiaries of the movement. The opening of Liebig's laboratory at the University of Giessen in 1824 was at once

of Mines (London, 1920), esp. 56–65. On the Royal College of Chemistry, see Gerrylynn K. Roberts. "The establishment of the Royal College of Chemistry: an investigation of the social context of early-Victorian chemistry," *Historical studies in the physical sciences, 7* (1976), 437–485.

57. Jean-Baptiste André Dumas, *Rapports adressés à M. le Ministre de l'Instruction publique* (Paris, 1846), 2–11. On Dumas's proposals and their consequences, see Robert Fox, "L'attitude des professeurs des facultés des sciences face à l'industrialisation en France entre 1850 et 1914," in Christophe Charle and Régine Ferré, eds., *Le personnel de l'enseignement supérieur en France aux XIXe et XXe siècles* (Paris, 1985), 135–147, on 136–141.

58. Both the general history of German higher technical education and the particular significance of the founding of the Polytechnische Schule of Karlsruhe are well treated in Franz Schnabel, "Die Anfänge des technischen Hochschulwesens," in *Festschrift des 100 jährigen Bestehens der Technische Hochschule Fridericiana zu Karlsruhe* (Karlsruhe, 1925), 1–44.

an indicator of what was afoot and a main inspiration for the chemical laboratories that were established in a number of other universities and in the more ambitious technical schools in the 1840s and 1850s (see fig. 1).[59] In this proliferation, which drew additional strength from a spirit of emulation fostered by the political fragmentation of Germany and the resulting rivalry between states, an enlarged conception of the purpose of the chemical laboratory was established. Chemical laboratories now came to be regarded as the location both for research and for the inculcation of new applicable skills, and, as such, they quickly assumed the status of essential accouterments of any modern university.

Baden was typical of several German states in which the provision of laboratory instruction in chemistry moved to a significantly higher level in the wake of Liebig's success. The new chemical laboratory of the Polytechnische Schule at Karlsruhe, opened in 1851 at a cost of £5000, and that of the University of Heidelberg, built between 1853 and 1855 for the newly appointed Robert Bunsen at a cost of £7300, were both far more expensive than the state government had bargained for.[60] But they justified what many saw as startling lavishness by the emphasis they placed on the applications of chemistry, in particular to agriculture—an emphasis that did not impede the establishment at Heidelberg of an internationally celebrated school of chemical research under Bunsen's charismatic leadership.[61] A similar orientation was apparent in other universities, where courses of the kind that Friedrich Knapp offered as extraordinary professor of technology and chemistry at Giessen from the 1840s began to multiply.[62]

A particularly notable innovation of this period was the creation of a new chair of agricultural chemistry and technology at the University of Würzburg in 1848. Johann Rudolf von Wagner moved to this chair, from

59. Brock, *Justus von Liebig* (ref. 36), 37–71, where the circumstances of the opening of the laboratory and the central role of organic analysis are examined in detail. On the importance of the Giessen laboratory as the setting for Liebig's research school, see Morrell's "The chemist breeders" (ref. 47).

60. Peter Borscheid, *Naturwissenschaft, Staat und Industrie in Baden (1848–1914)* (Stuttgart, 1977), 50–70. Bunsen's laboratory is well described in Theodor Curtius and Johannes Rissom, eds., *Geschichte der chemischen Universitäts-Laboratoriums zu Heidelberg seit der Gründung durch Bunsen* (Heidelberg, 1908), 4–26.

61. The sense of excitement that prevailed in Bunsen's laboratory is well described in *The life & experiences of Sir Henry Enfield Roscoe, D.C.L., LL.D., F.R.S. written by himself* (London, 1906), 54–57.

62. It was while teaching at Giessen, as *ausserordentliche Professor*, that Knapp published his important *Lehrbuch der chemischen Technologie, zum Unterricht und Selbststudium* (2 vols., Braunschweig, 1848–1853). See note 45 in this chapter, for the English translation.

the town's Gewerbeschule, following the death of the first holder in 1855, to begin a distinguished career as a teacher of applied chemistry, the founder and editor of the important *Jahres-Bericht über die Fortschritte der chemischen Technologie* (published annually for over eighty years from 1856), and the author of one of the most successful of all nineteenth-century textbooks in the field, his *Handbuch der chemischen Technologie* (entitled in earlier editions *Die chemische Technologie*), which went through fourteen German editions between 1850 and 1893 and numerous editions in English and French.

Despite the fuss and investment that accompanied their foundation, the new institutions and courses could only hope to achieve lasting success if they provided a type of education that met the expectations of employers and of students in search of economic advantage or professional advancement. There is evidence that these expectations were met in the German-speaking parts of central Europe, especially in the advanced polytechnic schools that were eventually to become Technische Hochschulen.[63] In Vienna in the mid-1840s, lectures on industrial chemistry were followed by over 90 students, of whom over a third were examined in the subject.[64] From the 1850s, as special curricula for technical chemists proliferated in the polytechnic sector, the trend to acceptance of this kind of instruction became even more marked. The substantial numbers studying chemistry, with a strong element of industrial chemistry, at Berlin, Karlsruhe, Stuttgart, and (especially after the opening of analytical and technical laboratories at the local canton's expense in the early 1860s) Zurich, provide striking evidence of a trend that gained strength once the region recovered from the political disturbances of 1848.[65]

It is a mark of the incipient divergence in economic performance that in the same period the comparable initiatives in France took root less successfully. Among the professors in the new science faculties of the 1850s, Louis Pasteur at Lille was exceptional for his capacity to put his "purist" formation at the Ecole Normale Supérieure at least temporarily behind him. Arriving at Lille as the faculty's first dean and professor of chemistry in 1854, he quickly adjusted his research interests and struck alliances with local industry by undertaking commissioned research and making visits to fac-

63. Ernst Homburg, *Van beroep "Chemiker." De opkomst van de industriële chemicus en het polytechnische onderwijs in Duitsland (1790–1850)* (Delft, 1993), 287–339 and 389–431.

64. Ibid., 392.

65. Ibid., 392–395. On the courses and facilities for chemistry at the ETH in Zurich, see *Bericht über die Organisation und das Wirken der Eidgenössischen Polytechnischen Schule in Zürich ausgearbeitet für die Welt-Ausstellung in Wien 1873* (Zurich, 1873), esp. 58–60 and 62–63.

FIG 1. Justus von Liebig's laboratory at the University of Giessen, opened in 1824, attracted students from within and beyond Germany and became the model of a modern university-based laboratory. From J. P. Hofmann, *Acht Tafeln zur Beschreibung des chemischen Laboratorium zu Giessen* (Heidelberg, 1842). Among those on the picture who have been identified are: [standing center, facing left] Heinrich Will (Leibig's successor at Giessen), [standing center right, back to the artist] A.F.L. Strecker (later professor of chemistry at Würzburg), [standing right, facing the bench] Johann Josef Scherer (later professor of medicine at Würzburg), and [extreme right] A.W. Hofmann. Liebig's laboratory assistant, "der dicke Aubel," is seated in the middle. Reproduced with permission of the Deutsches Museum, Munich.

tories a routine part of the curriculum.[66] But in other faculties, possibly the majority, professors found it hard to break with the *cursus* of the university system in which they had been fashioned, and their overtures to regional economic interests often lacked substance or even conviction.[67] More crucially, enrollments for applied instruction remained low. For example, in the face of persistently paltry numbers entering for the new intermediate qualification of the *certificat des sciences appliquées*, the courses that had been established to prepare for it in both the *écoles préparatoires à l'enseignement des sciences et des lettres* and the faculties were gradually abandoned through the 1860s or allowed to revert to a more traditional form of scientific instruction. Industrialists, for their part, continued to look for employees with a more modest and narrow technical formation than the reforms of the 1850s had been intended to provide; hence their loyalty to apprenticeship or more informal procedures for learning by doing remained essentially undimmed.

The demands of British industrialists too were overwhelmingly directed to the lower and middle levels of the labor market, and employers who regarded training on the job as a satisfactory norm remained the majority. As a result, while the gap between the academic world and that of manufacturing narrowed in Britain about the mid-century, it did so with demoralizing slowness, leaving the contexts in which science might become more easily integrated with practice disappointingly rare. To this extent, the campaigns of Playfair, De la Beche, Percy, and others in favor of a science-based industrial training must be regarded as only partially successful: as William Siemens observed in his presidential address to the Iron and Steel Institute in London in 1877, the nations of continental Europe had led Britain in their recognition of the value of technical education, and it was only increasing competition that had forced British manufacturers to open their minds tardily to the potential of a form of instruction founded on "those natural sciences which bear upon practice" (practice itself being something to be taught "in the workshop and in the metallurgical establishment").[68]

The same qualification about slowness applies to other fields as well. Take, for example, the signs of a growth of activity in the theoretical foun-

66. Denise Wrotnowska, *Louis Pasteur. Professeur et doyen de la Faculté des Sciences de Lille (1854–1857)* (Paris, 1975), esp. 55–56 and 70–75.

67. Fox, "L'attitude des professeurs" (ref. 57), 139–141 and (on the perpetuation of occasional signs of a similar reticence in Clermont-Ferrand on the eve of the first world war) 142–143.

68. Charles William Siemens, "President's inaugural address," *The journal of the Iron & Steel Institute*, 1877, 6-34, on 7-9. The Iron and Steel Institute had been founded in 1869 precisely to advance the ideals of "a judicious combination of natural science with practical experience" (ibid., 6) to which Siemens aspired.

dations of mechanical engineering—signs that became real during the 1860s and 1870s. Here, the popularity of William Rankine's *Manual of applied mechanics* (1858) and *Manual of civil engineering* (1861) as texts for engineering students in Britain and elsewhere was a mark of an incipient change of attitudes. Yet it was some years before the combination of Rankine's intellectual stature and rhetorical dexterity in support of his vision of a "harmony of theory and practice" evoked a recognizable response among the employers of academically educated engineers of the kind he prepared as Regius Professor of Civil Engineering and Mechanics at the University of Glasgow from 1855 to 1872.[69]

It is beyond question that in textiles, metallurgy, and mechanical engineering, British industry was generally resistant to change. Yet even as dyestuffs manufacturers, dyers, and calico-printers continued to employ colorists in ones and twos and in laboratories of legendary simplicity and untidiness, the intellectual equipment of the senior technical staff entering industrial chemistry through the 1850s and 1860s underwent a cautious but unmistakable adjustment. The case of the engagement of Playfair at Clitheroe was unusual, even unique, in 1841. But over the next quarter of a century home-trained products of the Royal College of Chemistry and University College in London and Owens College in Manchester, and others who had passed through the lofty mill of research training at a German university, began to tread in the footsteps of Playfair, Richardson, and Ronalds and constitute a dispersed but recognizable network of chemists who had passed from an advanced academic education to occupy senior industrial positions.[70]

69. Rankine's *Manual of applied mechanics* and his *Manual of civil engineering* drew on both the scientific education he received at the University of Edinburgh and his experience, over more than fifteen years, as a railway engineer. While rigorous in his treatment of the theoretical foundations of the problems he treated in his books, Rankine also offered detailed practical illustrations of the problems under discussion. On Rankine's career and views on engineering education, see P. G. Tait's biographical essay in William John Millar, ed., *Miscellaneous scientific papers of W. J. Macquorn Rankine* (London, 1881), xix–xxxvi; David F. Channell, "The harmony of theory and practice: the engineering science of W. J. M. Rankine," *Technology and culture, 29* (1988), 98–103; and Ben Marsden, "Engineering science in Glasgow: economy, efficiency and measurement as prime movers in the differentiation of an academic discipline," *The British journal for the history of science, 25* (1992), 319–346. For a clear statement by Rankine himself, see the inaugural lecture he delivered at the University of Glasgow in 1856: *Introductory lecture on the harmony of theory and practice in mechanics* (London, 1856).
70. Bud and Roberts, *Science versus practice* (ref. 23), esp. 97–118. As Bud and Roberts stress, however, a high proportion of those who attended the English schools we mention had done so as "occasional" students, following individual courses or laboratory classes, rather than full programs of study.

While the diffusion of formally qualified chemists into industry, in particular of those with some experience of laboratory work, had an inevitable impact on attitudes and procedures in the firms they entered, it is still important to stress both the small numbers involved and the difficulty of the process of integration. As John Mercer's comments remind us,[71] the suspicion that the older employees in the chemical industry expressed towards the work of the newcomers reflected not only an understandable rivalry but also a just perception that the conception of research and the fine bench technique that they brought with them from their experience of a university laboratory were hard to realize in the context of contemporary industrial technology. A modern-minded British employer in, say, 1860 might regard the skills of a trained chemist as a means of fostering efficiency or enhancing quality. But he was unlikely to see any need for the kind of restructuring and adjustment of objectives that would have given a chemist trained in research the opportunity of deploying the full range of his abilities. In particular, he would not see the duties of his chemist extending to the development of fundamentally new products or processes.

New opportunities: Chemical dyestuffs

The preparation of the first fully synthetic dye, mauve, by William Henry Perkin in 1856 is a potential historiographical snare. Since Perkin discovered the dye while working (on the synthesis of quinine) in Hofmann's laboratory at the Royal College of Chemistry, the event is easily taken as a watershed in the history of applied research, a decisive first step on the path to the modern in-house laboratory, with its stress on science-based innovation, that culminated in the huge laboratories of the Badische Anilin- und Soda-Fabrik (BASF), Bayer, and General Electric half a century later.[72] In fact, the contemporary evidence underlines not so much a transition to a new age but rather the essential continuity between the structures for the assimilation of science in the chemical industry before and after Perkin's work. The explanation for this lies, above all, in the nature of the technol-

71. See above, 17.

72. In the plentiful secondary literature on Perkin's discovery and the early history of synthetic dyestuffs, see, for example, L.F. Haber, *The chemical industry during the nineteenth century. A study of the economic aspect of applied chemistry in Europe and North America* (Oxford, 1958), 81–87; John Joseph Beer, *The emergence of the German dye industry* [Illinois Studies in the Social Sciences, vol. 44] (Urbana, IL, 1959); and Anthony S. Travis, *The rainbow makers. The origins of the synthetic dyestuffs industry in Western Europe* (Bethlehem, PA, 1992). A rich source, with a geographical coverage rather greater than its title would suggest, is George F. Jaubert, *Historique de l'industrie suisse des matières colorantes artificielles* (Geneva, 1896).

ogy for the production of the new generation of aniline dyes of which mauve was the precursor. An essential point is that the dyes that could be made, all of them derived from coal tar, were not numerous: those that enjoyed real commercial success were variants on no more than about six basic types.[73] Moreover, their invention and development depended less on precise, refined apparatus or deft experimental technique than on practical skill and know-how. As a result, the laboratories of the dozen or so companies that were manufacturing aniline dyes in Britain, France, Germany, and Switzerland by 1861 retained the character of workshops where trial and error, although probably more systematic than in the heyday of natural dyes, remained the core of the chemist's practice.

It was a natural corollary of this continuity that the early companies remained small, and that, far from setting out as a matter of policy to broaden their repertoire of products, they were normally content to concentrate on one artificial dye or, at the most, on a very limited range of dyes which they regarded as their specialities in the field. Commonly a dye would be invented by an individual working outside or in loose association with the industrial context; it would then be bought and exploited by a manufacturing company, which would often maintain its interest in natural dyes as well. The case of Perkin, who made his discovery in an academic environment and began to exploit it in the small company he founded with his father, broadly squares with this pattern. It is also entirely characteristic of the early history of aniline dyes that while Perkin went on to pursue research in his private laboratory, his company made no attempt to use mauve as the starting point for the development of a wider range of colors. The company lived by mauve alone and it was sold in 1873 as an enterprise whose production was dominated by this one dye.

The case of another early aniline dye, variously known as magenta, aniline red, or fuchsine, is similar. In France, the patent covering Emmanuel Verguin's original method of preparation was taken out in 1859 by the Renard brothers, the owners of a silk-dyeing firm in Lyon, and it passed four years later, following mergers and a complex restructuring, to a specially created company, the Société La Fuchsine.[74] Although the Société La Fuchsine was founded to manufacture and sell a wide range of products

73. For an excellent contemporary account of the groups of aniline dyes, see Hofmann's report on Class II, Section A, of the International Exhibition of 1862 (ref. 40), 119–132.

74. On the Société La Fuchsine and the litigation concerning the dye, see, in addition to the more general sources cited in note 72, Henk Van den Belt, "Action at a distance: A.W. Hofmann and the French patent disputes about aniline red (1860–63), or How a scientist may influence legal decisions without appearing in court," in Roger Smith and Brian Wynne, eds., *Expert evidence. Interpreting science in the law* (London, 1989), 185–209, and "Why monopoly failed: the rise and fall of Société La Fuchsine," *The British journal for the history of science, 25* (1992), 45–63.

from coal tar, it devoted most of its energy to the protection of its small core of main patents, in particular the one for aniline red, and did nothing to increase the number of dyes it produced. In Britain, the somewhat different method of preparation developed by Henry Medlock and independently by the French chemists Charles Girard and Georges De Laire was bought and exploited by the London manufacturers Simpson, Maule, and Nicholson. But here again, aniline red entered production as one of a small, stable range of speciality dyes rather than as the starting point for the creation of new colors.

Despite the essentially quiet assimilation of the early aniline dyes within industry, the first decade of their history came gradually to be marked by trends that discretely encouraged a more sophisticated approach to the study of color chemistry. The trends were less evident in the world of practice, in the simple workshops of dyers and colorists, than in the expanding world of universities and schools of advanced technical education. Hofmann's initial interest was typical in that it seems to have grown from the purely intellectual challenge of a family of synthetic products of complex and uncertain composition.[75] But soon curiosity and the understanding that resulted from it found a practical outlet in the patent litigation that moved abruptly to center stage in the early 1860s. No case was more notorious than the series of disputes in which the Société La Fuchsine struggled to defend its right against rivals who had devised other methods of producing aniline red. Suddenly, and quite unequivocally, the exact techniques of analysis that were the stock in trade of academically trained chemists had found an economic function. Jules Persoz, Emil Kopp, and (most prominently) Hofmann were just the best-known of the chemists, most of them working outside industry, who weighed in with far from unanimous views on the distinctions that were to be drawn between the various forms of fuchsine and similar dyes and the processes by which they were manufactured.

In advancing chemists to the rank of acknowledged, uniquely qualified expert witnesses, the fuchsine affair contributed significantly to the acceptance of laboratory science as an essential part of industrial activity rather than a self-indulgent adjunct of uncertain relevance. One mark of acceptance was the increased flow of books and papers by academic chemists working on the new dyes. The textbook of chemical dyeing, with the pano-

75. Anthony S. Travis, "Science's powerful companion: A. W. Hofmann's investigation of aniline red and its derivatives," *The British journal for the history of science*, *25* (1992), 27–44.

ply of complex equations, fine analyses, and pasted-in samples of dyed material—all very different from an earlier generation of narrowly practical manuals[76]—became a regular part of the organic chemist's literary production, along with research papers that were soon staple fare in the relevant learned journals throughout Europe. And soon no country lacked a detailed and authoritative text on the chemistry of synthetic dyestuffs. Sometimes, as in the case of Paul Schützenberger's *Traité des matières colorantes*, the texts would include a parallel treatment of natural dyes;[77] but more commonly, as with *Die Theerfarbstoffe* of P.A. Bolley and Emil Kopp, they were devoted almost exclusively to the new dyes.[78]

Despite the scientific sophistication of the literature in journals and books, there remained, even in the late sixties, marked divergences between the work of the chemists in the dye factory and the printing works, oriented to modest improvement and rather imprecise quality control, and that of the academic commentator, with its emphasis on the fine distinctions between different products and methods of production and on the underlying chemical principles. Even if the academic chemist was glad to intervene, in a law suit or in suggesting improved methods or cheaper raw materials, his interventions still tended to be occasional rather than part of the day-to-day functions of the factory and to come from outside the industrial context. It was only with the synthesis of the first alizarin dye by the German chemists Carl Graebe and Carl Liebermann in 1869 and, more conspicuously, the development of the azo dyes from the 1880s that the gap between the research-oriented chemistry of the academic or private laboratory and the more mundane, economically driven activities of the production plant began to close conspicuously. The families of the alizarin and azo dyes were potentially far larger than that of the aniline dyes, so that a high degree of product differentiation and an unprecedented competitiveness in the quest for new patents and products now became leading characteristics of the dyestuffs industry. Equally marked was the move to greater analytical precision and hence to a sophistication in facilities and laboratory skills that would have been out of place in the aniline phase. We shall have more to say about this in our comments on the promotion of systematic in-house research in Chapter 4.

76. For a typical British product of this generation, see Edward Andrew Parnell, *Dyeing and calico-printing* (London, 1849); reprinted from the same author's *Applied chemistry in manufactures, arts, and domestic economy* (London, 1844).
77. Paul Schützenberger, *Traité des matières colorantes* (2 vols., Paris, 1867).
78. Pompejus Alexander Bolley and Emil Kopp, *Die Theerfarbstoffe* (3 vols., Braunschweig, 1867–1897). On the circumstances that caused the publication of this book to be so protracted and in the end to be the work of several authors, see Richard Meyer's introduction of April 1897 to the reissued first volume, v–vii.

New opportunities: Precise measurement in thermal physics and telegraphy

In the mid-nineteenth century, at about the same time as precise laboratory techniques were acquiring greater prominence in both the pedagogy and the practices of chemistry, standards of precision in laboratory physics were also being raised in contexts that owed much to the changing profile of industrial demands and opportunities. As in the case of chemistry, there were long-established precedents. These were to be found most obviously, though by no means exclusively, in a French tradition of fine experimenting that had yielded, in electricity, Coulomb's systematic examination of electrostatic forces in the later eighteenth century and, a few years later, in optics, the work of Biot, Malus, and Fresnel.[79] But it was in the study of thermal phenomena that French physics about the mid-century made its greatest mark. In this field, too, the French had been prominent since the first quarter of the century. Then, measurements of the specific heats of gases and solids had been a speciality and a source of influential data, including not only the measurements of specific heats that had led to the establishment of Dulong and Petit's law but also the false and seriously misleading observation of Delaroche and Bérard that the specific heat of air decreased when air was compressed.[80] The main architect of the renewal of interest in work in this area was Victor Regnault, a former pupil of the Ecole Polytechnique and Ecole des Mines who had specialized in chemistry (passing through the laboratories of Liebig at Giessen and the agricultural chemist Jean-Baptiste Boussingault at Lyon) and who first assisted and then, from 1840, succeeded Gay-Lussac in the chair of chemistry in Paris at the ancient and very prestigious research institution, the Collège de France.

From his disciplinary roots in chemistry, Regnault turned, at about the time of his appointment to the Collège de France, to the experiments on the thermal properties of matter in all three states, though especially of gases and vapors, for which over the next thirty years he was to become famous.[81] Patronage for the experiments came, in generous quantities, from

79. John L. Heilbron, *Electricity in the 17th and 18th centuries* (Berkeley, 1979), esp. chapter 19, and Jed Z. Buchwald, *The rise of the wave theory of light. Optical theory and experiment in the early nineteenth century* (Chicago, 1987).
80. Robert Fox, "The background to the discovery of Dulong and Petit's law," *The British journal for the history of science*, 4 (1968–1969), 1–22, and *The caloric theory of gases from Lavoisier to Regnault* (Oxford, 1971), 131–142 and 238–248.
81. Regnault's results were published as *Relation des expériences entreprises par ordre de Monsieur le Ministre des Travaux Publics . . . pour déterminer les principales lois et données numériques qui entrent dans le calcul des machines à vapeur* (3 vols., Paris, 1847–

a committee of the Ministry of Public Works concerned primarily with the accumulation of accurate data relevant to the safety and efficiency of steam engines. Regnault's earliest work was directed to the determination, to a standard of exactness unmatched in any earlier experiments, of the variation in the pressure of steam and other vapors with temperature and of the specific heats of vapors and gases. The data on pressure were particularly important in the continuing struggle to prevent explosions of stationary and marine steam engines, and both these data and those on specific heats contributed to a better understanding of the parameters that affected the consumption of fuel, especially in the new generation of medium- and high-pressure engines that were coming into widespread use in the mid-nineteenth century. In fact, Regnault's observations achieved a precision greater than was necessary for the practical demands of steam-power technology that stimulated his research, but the results immediately became standard ingredients of engineers' manuals and remained so for many decades. They also acquired prominence among physicists through their application in the papers of Rudolf Clausius and William Thomson that laid the foundations of modern thermodynamics.[82]

The quality of Regnault's observations earned an international reputation for the laboratories in which he worked, initially at the Collège de France and then, from 1854, at the state porcelain factory at Sèvres, where he was director. It also set a daunting and very visible standard for good experimental physics about the mid-century. William Thomson's decision to gain experience by working with Regnault in 1845 just after his graduation from Cambridge took him to what was then the most prestigious of all sites for exact physical measurement.[83] He went with the immediate, limited aim of strengthening his qualifications, in particular as a teacher, for the chair of natural philosophy at the University of Glasgow to which he aspired. But he gained far more than this. As a physicist, he saw at first hand the painstaking approach that set Regnault head and shoulders above his peers in experimental physics: in later life, he was to recall the vision he had absorbed of a "faultless technique, a love of precision in all things, and the highest virtue of the experimenter—patience."[84] And no less crucially, his close association with Regnault provided a visible demonstra-

1870); also published, with identical pagination, as *Mémoires de l'Académie [Royale, Impériale] des Sciences de l'Institut de France, 21* (1847), 1–767; *26* (1862), iii–x and 3–928; *37:1* (1868), 3–575, and *37:2* (1870), 579–968.

82. Donald S.L. Cardwell, *From Watt to Clausius. The rise of thermodynamics in the early industrial age* (London, 1971), 239–294.

83. Crosbie Smith and M. Norton Wise, *Energy and empire. A biographical study of Lord Kelvin* (Cambridge, 1989), 104–108.

84. Ibid., 108.

tion that, far from being mutually exclusive, utility and precision physics shared much common ground.

Regnault's easy moves between the concerns of the laboratory and those of practical power technology made a natural impression on a young man who had been brought up, from the age of eight, in the booming manufacturing city of Glasgow and in close association with an elder brother who went on to become professor of civil engineering at Queen's College, Belfast, and then professor of engineering at Glasgow. Thomson's own career was to maintain a versatility similar to Regnault's. Characteristically, he performed some of the most important of his experiments on gases and heat in the industrial environment of Salford with James Prescott Joule, the son of a prosperous family of brewers, in the late 1850s.[85] And his sensitivity to the applications of physics helped to draw him, from 1854, into a concern with telegraphy, one of the classic settings in which science and engineering interacted and overlapped with each other during the second half of the nineteenth century.

The electrical telegraph was to remain an abiding interest throughout the rest of Thomson's creative years as a physicist and to stimulate some of his most important research. Since the pioneering work of Samuel Morse in the U.S. and Wheatstone in Britain in the 1830s, the technology had made great strides, and by the late 1840s, the construction of long-distance underground and submarine lines, made possible by the introduction of gutta-percha as an insulating material in 1848, had raised it to new levels of sophistication, creating problems that it became increasingly difficult for telegraph engineers to solve without drawing on the theoretical and experimental skills of physicists. It was an investigation of these problems by Michael Faraday, following the mixed success of early cables between Dover and Calais and across the Irish Sea, that first attracted Thomson's attention.[86] Apart from the difficulty of laying a cable strong enough to withstand the mechanical impact of currents and damage from anchors, the main concern of telegraph operators was the retardation and weakening of the signal, an effect that made rapid, accurate transmission over long distances impossible.

Before the Electric Telegraph Company consulted Faraday, companies had relied on the rule-of-thumb advice of men who had gained most of their know-how in the field—in the construction of overground or underground land-lines, over short or medium distances—and whose expertise, though adequate to most of the challenges of land-lines, was glaringly

85. Ibid., 301–316, and Donald S.L. Cardwell, *James Joule. A biography* (Manchester, 1989), 122–155.
86. Smith and Wise, *Energy and empire* (ref. 83), 446.

wanting in the new age of submarine telegraphy.[87] Faraday's approach was very different. With the perspective of a physical scientist, he offered an analysis that identified the source of the problem in the huge inductive capacity of cables submerged in water.[88] He could offer no solution, however, and it was left to younger and mathematically more competent physicists to explain the retardation and diminished quality of the signals. Thomson, who devised a mathematical treatment that drew on his reading of Fourier's analysis of the conduction of heat, was prominent among those who responded to the challenge. As he became more deeply engaged in practical telegraphy, following his appointment as a Director of the Atlantic Telegraph Company in 1856 and his involvement in preparations for the laying of the first (unsuccessful) transatlantic cable in 1858, Thomson's attention turned increasingly to the deleterious consequences of the use of copper of uneven quality as the core of the cables.[89] The precaution that he advocated was the adoption of rigorous experimental procedures for the determination of the resistivity of the copper and of the properties of the insulating materials that affected the inductive capacity of the cables.

To achieve the degree of precision that came now to be considered necessary, new laboratory equipment and accurate, reproducible electrical standards—and, even more importantly, a new generation of men able to use them—were required. The resulting stimulus was felt throughout the discipline of physics. In this phase, Thomson's sensitivity to the scientific and economic value of precision earned him both reputation and material prosperity. His marine mirror-galvanometer, an instrument capable of detecting weak signals, was devised for use with the transatlantic cable of 1858; and three years later he played a central role in the establishment of a committee of the British Association for the Advancement of Science committee appointed to define an absolute unit of resistance and to design an easily reproducible standard that would represent the unit.[90] Over the next nine years, until the end of the first phase of the project in 1870, Thomson remained a leading figure among the experimental physicists who pursued the complex and often controversial tasks set by the BAAS commit-

87. For a good account of the changes in the nature of the technical problems that followed the advent of submarine telegraphy, see Bruce J. Hunt, "The ohm is where the art is: British telegraph engineers and the development of electrical standards," *Osiris*, 2nd ser., 9 (1994), 48-63.

88. On Faraday's analysis and the response of Thomson, Stokes, and Maxwell to it and the problem of retardation, see Smith and Wise, *Energy and empire* (ref. 83), 445–494.

89. Ibid., 661–683.

90. Ibid., 669–672 and 687–698. On the BAAS initiative and its increasingly international context, see also Simon Schaffer, "Accurate measurement is an English science," in M. Norton Wise, ed., *The values of precision* (Princeton, NJ, 1995), 135–172.

FIG 2. The siphon recorder, which William Thomson developed to replace his minor galvanometer and spark recorder for the reception of signals transmitted through submarine cables. Thomson obtained his first patent for the siphon recorder, in which signals were recorded automatically on a moving paper tape, in 1867. This model shown in the plate is an improved one that he described in 1873, and further refinements continued to be made into the 1880s. From William Thomson, "On signalling through submarine cables," in Thomson, *Mathematical and physical papers* (6 vols., Cambridge, 1882-1911), 2, 1678-172 (170), originally published in *Transactions of the Institution of Engineers and Shipbuilders in Scotland*, 18 Mar 1873. Reproduced with permission of the London Library.

tee. And he maintained his interest when the committee's mission was reactivated, in the context of growing international interest, in 1880.

Electrometers of the divided ring and quadrant type for the measurement of potential difference and the siphon recorder that he patented in 1867 for the automatic recording of signals (see fig. 2) were just three of the best-known products of Thomson's work. No less important, however, was his day-to-day demonstration that precise measurement and the determination of absolute units and standards stood at the cutting edge of physics, both pure and applied. As Thomson knew very well, his simultaneous involvement in business, experimental work of the kind required by the BAAS committee, and the more rough and ready outdoor world of telegraphy helped to provide both the intellectual stimulus of a coherent programme and the resources he needed for the continuing improvement of his laboratory. Especially after the laying of the first successful transatlantic cable in 1866, the strategy yielded rich dividends. Patents were profitable, though relentlessly time-consuming, and agreements with various telegraph companies from the late 1860s opened the way to substantial consultancy fees.[91] An income from patents and fees that was already considerable in the 1860s and amounted to several thousand pounds a year by about 1880 provided an ample source of funds for research and for the characteristic style of research-oriented teaching that Thomson developed to ever greater effect, first in the cramped quarters of the old Glasgow college and then in the new buildings to which the university moved in 1870.

The repercussions that the quest for standards and precision had on the practice of laboratory physics were immediate and profound in Britain, as they were in all industrialized countries.[92] As was quickly recognized, an area of application in which trained physicists possessed an expertise quite distinct from anything that contemporary engineers could offer had been fashioned. Moreover, it was now beyond question that the acquisition of laboratory skills of a kind that soon came to be valued by the telegraphic and other industries could at the same time be scientifically respectable. Spurred on by this incentive, physics followed a path already taken by chemistry, by establishing procedures, analogous to those of chemical analysis, that were easy to use with a large class and, once the expenditure on the laboratory and equipment had been made, inexpensive.

In recognition of the benefits of bringing together the academic and industrial worlds, Thomson, though pre-eminent in his visibility, was by no means alone. As we shall see in Chapter 2, many other practitioners of physics at about the same time independently perceived precisely the opportunities that he exploited. The fruits quickly appeared, initially in a new confidence that helped to engender a keen sense of the material needs of the discipline and then in the conspicuously improved laboratories that resulted from that sense, especially after 1870. Simultaneously, and crucially for the subject of this book, the kinds of pedagogical initiatives that Thomson and other physicists introduced during the third quarter of the nineteenth century helped to provide both the material foundations and the intellectual impetus for the emergence of physics in general, and of electricity in particular, as a leading area of scientific research which no self-respecting university could afford to neglect.

91. Smith and Wise, *Energy and empire* (ref. 83), 698–712.
92. On the repercussions in Britain, see Romualdas Sviedrys, "The rise of physics laboratories in Britain," *Historical studies in the physical and biological sciences, 7* (1976), 405–436, and Graeme Gooday, "Precision measurement and the genesis of physics teaching laboratories in Victorian Britain," *The British journal for the history of science, 23* (1990), 25–51.

The concepts and practices of what may broadly be defined as applied research began to make their mark in some of the main industrialized countries of Europe in the early and mid-nineteenth century. Such research sometimes found a home within factories, for example in the modest color shops of the kind that the self-taught John Mercer directed in Lancashire. But, increasingly, applied research attracted the attention of academic scientists who, like Hofmann, Regnault, and William Thomson, moved easily between the realms of industry and higher education and even used the facilities of their institutions in pursuit of their commercial interests. By the end of the century, the implantation of industrially oriented research in universities and advanced technical schools had become so widespread that the obstacles to what had occurred were, and still are, easily overlooked. One leading impediment was the slowness with which research of any kind, with or without industrial implications, came to be accepted as an essential function of higher education. It was only from the mid-century, in fact, that the institutionalization of research began to accelerate. The process that ensued can be traced in almost any country, and, as the growth of the great American research universities of the late nine-teenth century shows with particular clarity, it was by no means an exclusively European phenomenon or the affair of a particular discipline. In this chapter, however, we narrow our focus to three national contexts—those of Germany, France, and Britain—and to the case of physics.

As we argue, research became formalized as a central concern of academic physicists in the context of a virtually ubiquitous parallel movement that fostered novel approaches to laboratory-based pedagogy after 1850. In the process, teaching and research fed off each other: as the needs of laboratory instruction elicited ever greater support, research too was commonly the beneficiary. One consequence of the movement was that professors found themselves engaged in a relentless quest for space and equipment. The intensity of the quest differed from country to country, and there was much unevenness in the capacity of the various national systems of higher education to respond at a level that would allow the conversion of the new ideals into reality. There were differences, too, in the ease with which laboratory teaching and research managed to coexist, although everywhere, despite some inevitable tensions, the interests of the discipline meant that a *modus vivendi* had to be found.

The difficulties arising from the proximity of teaching and research usually stemmed from the priority that was almost invariably given to the needs of the former. The leading role of teaching as a motor for the proliferation of academic laboratories from the mid-century, which was quite as marked in the biological sciences as it was in the physical sciences, has been explored in detail in the secondary literature, and we shall not rehearse all the facets of what is now a familiar account. Instead, in the present chapter, as in the one that follows on electrical engineering, we take a particular perspective, to stress the extent to which the increasingly powerful ideals of laboratory-based pedagogy and research affected and were affected by the unprecedented emphasis on the industrial applications of science.

In the simultaneity of the changed priorities of academic life and of the new openness of science to industry, there lay yet another recipe for tension. On the one hand, as we argue here with respect to physics, the applied dimensions of the discipline, as of other sciences, were recognized to be of vital importance in enhancing the revenues that universities and higher technical schools needed for the creation of well-founded modern laboratories. On the other hand, however, as the funds were expended and facilities improved, the pride and disciplinary loyalties of teachers of physics fed a new sense of the dignity of their calling as the creators of knowledge and the exemplars of a fiercely independent academic profession committed to the liberal spirit of research and learning. On this score too, the balance between the demands of utility and those of disinterested inquiry was never easily struck. Debates about it and attempts to maintain it remained leit-motivs of the whole period from the mid-nineteenth century to the first world war, and they provide a recurring theme of this chapter.

Germany: Research and the ideals of the university

The lengthy process that brought research in physics to the status of a fully institutionalized activity of higher education began to assume new momentum in the late 1840s and to do so most markedly in the German states. Explanations of what occurred there—in all disciplines and not just in physics—have tended to turn on two main factors.

The first factor was the reform movement that, between 1806 and 1818, transformed the spirit and administrative structure of higher education in Prussia, after defeat at the hands of Napoleon's army. At the heart of the reform lay the ideal of *Bildung*, a process of cultivation through intellectual endeavor that is associated indissolubly with the name of the Prussian Minister of Education, Wilhelm von Humboldt. With their stress on knowledge pursued for its own sake by rigorous systematic inquiry, Humboldt's objectives broke sharply with the old traditions of professional training for careers in government service or in law, medicine, or the Church. In doing so, they encouraged the spread of higher intellectual standards among the professoriate and the mass of the student population, notably the very substantial proportion who aspired to careers in teaching in secondary schools. One corollary of the raising of standards was a growth of interest in the acquisition and diffusion of new knowledge and in the conditions that these activities required. Almost immediately the research seminar became a leading element in the Humboldtian university, building on roots in the seminars of classical philology that had existed as early as the 1730s at Göttingen. In the traditional classical seminar, research took the form of a critical analysis of ancient authors by a small group of scholars and advanced students, and novelty consisted in the production of philologically sound texts and in a better understanding of them.[1] As comparable objectives made their mark in other disciplines, including the natural sciences and mathematics (from the 1820s), they necessarily assumed different forms.[2] But, whatever discipline was involved, the seminar or its scientific equivalent (usually modelled on Liebig's laboratory at Giessen) promoted the instruction of pupils through the imparting of a rigorous technique that was first learned by example and then applied, under supervision, to the advancement of knowledge.

The second, closely related factor that signalled and encouraged the integration of research in German universities was the growing professionalization and specialization of university careers.[3] One mark of the movement was a new emphasis on the formal intellectual qualifications required of professors and their assistants. The essential qualification was the doctorate of philosophy, for which candidates had to give at least some evidence of independent thought, demonstrated in an oral ex-

1. R. Steven Turner, "University reformers and professorial scholarship in Germany, 1760–1806," in Lawrence Stone, ed., *The university in society* (2 vols., Princeton, NJ, 1974), 2, 495–531. On the pedagogy of the eighteenth and early nineteenth centuries, see also the classic Friedrich Paulsen, *Die deutschen Universitäten und das Universitätsstudium* (Berlin, 1902), esp. 52–82, and the earlier and somewhat briefer account in Paulsen, *The German universities. Their character and historical development*, trans. Edward Delavan Perry (New York and London, 1895), 57–88.

2. R. Steven Turner, "The growth of professorial research in Prussia, 1818–1848—causes and context," *Historical studies in the physical sciences, 3* (1971), 137–182.

3. Although professionalization and specialization were both fostered by Humboldt's vision, specialization in particular was not always easily reconciled with the breadth of the intellectual engagement of the ideal Humboldtian professor. See the observation by R. Steven Turner, cited in below, p. 44.

amination and usually (though not invariably) the writing of a doctoral thesis. Typically, the doctorate was not a formidable hurdle: the thesis seldom called for work of fundamental originality, and the whole degree could be obtained by an able student in a matter of months (as it was by the physicist Arthur Schuster, at the age of 21).[4] Despite its modest character, doctoral research at first did not fit easily into the German university. The move to specialization that it represented sat uncomfortably with Humboldt's cultural ideals, which made as much of the unity of knowledge as they did of research, and for that reason, as well as for the threat it posed to university finances, many professors viewed it with suspicion.[5] By the mid-century, however, these tensions had eased, leaving advanced, specialized study an activity on which no German university of consequence could turn its back.

Of the sciences, physics had initially a less prominent place in these developments than the more established disciplines of medicine and chemistry. But in the middle decades of the century the champions of physics became more assertive. They sought not only to define the elevated academic credentials that would give the subject the prominence they believed it deserved in the world of learning, but also to establish institutional structures analogous to those in other disciplines that by now had come to set greater store by research. Among these other disciplines, chemistry was generally seen to provide the most relevant model. By 1850, in nearly all German universities, chemical laboratories had two main functions that aroused the interest of physicists. At one level, students of pharmacy and medicine were still trained in relatively routine manipulative skills, much as they had been in the early days of the Giessen laboratory. But also, and more importantly, the best chemical laboratories had come to provide a context in which new ideas and practices were fostered and in which professors systematically inspired and participated in the research of their pupils.

The process by which physics laboratories acquired a similar balance of activities was not complete until the 1870s. But already, by the 1840s, the discipline in Germany had made modest but crucial steps towards greater visibility and respectability. German-language journals, such as J.C. Poggendorf's *Annalen der Physik und Chemie,* no longer devoted the lion's share of their space to translations of papers originally published in En-

4. Arthur Schuster, *Biographical fragments* (London, 1932), 56–64. Henry Roscoe records that he passed the examinations for his doctorate at Heidelberg six months after his arrival; see *The life & experiences of Sir Henry Enfield Roscoe, D.C.L., LL.D., F.R.S. written by himself* (London and New York, 1906), 58.
5. Turner, "The growth of professorial research" (ref. 2), 154–156.

glish or French, as they had done earlier in the century. And German papers, for their part, were beginning to win more recognition abroad. From the mid-century, the number of German contributions to both German and foreign journals grew steadily, reflecting the mounting pressure to publish as, in physics as in other disciplines, evidence of activity in research and publication came to weigh more and more heavily in appointments to chairs and in the struggle for reputation that fired all but the least ambitious universities.

In the history of the discipline in Germany, Gustav Kirchhoff's appointment as professor of physics at the University of Heidelberg in 1854 stands as a landmark. Kirchhoff received his chair, first and foremost, for his record in original research; he occupied it at the early age of thirty and with the strong backing of the philosophical faculty, in which the chair was located.[6] The faculty's insistence on this criterion, rather than on Kirchhoff's record as a teacher, set the tone for his subsequent career at Heidelberg, and for the next twenty years his research, mainly on elasticity, electricity, spectrum analysis, and radiant heat, won him national and international fame and students. Among the young foreign physicists who worked at Heidelberg during his long reign were Gabriel Lippmann (from France), Kammerlingh Onnes (from Holland), Ludwig Boltzmann (from Austria), and (from Britain) Arthur Schuster.[7] All of them were attracted not by the laboratory, which remained small and poorly equipped even in the early 1870s,[8] but by the attitudes to research reflected in Kirchhoff's appointment and espoused more broadly at Heidelberg, notably by Kirchhoff's colleague in the chair of chemistry and collaborator in his work on spectrum analysis, Robert Bunsen.

The sense that Germany was forging ahead in physics and that a main key to her success lay in the criteria for professorial appointments emerged in the 1850s and 1860s as part of a more general international perception of the invigorating effect of research in German universities. It was a percep-

6. Christa Jungnickel and Russell McCormmach, *Intellectual mastery of nature. Theoretical physics from Ohm to Einstein* (2 vols., Chicago and London, 1986), *1*, 288–294.
7. Ibid., *1*, 293n. For Schuster's reminiscences of the year he spent working with Kirchhoff (in 1872–1873) and his contacts with Lippmann and Onnes, see Schuster, "Biographical byways," *Nature, 115* (1925), 126–127, *Biographical fragments* (ref. 4), 56–65 and 216–222, and *The progress of physics during 33 years (1875–1908). Four lectures delivered to the University of Calcutta during March 1908* (Cambridge, 1911), 13–15. Further personal testimony to the excitement that existed at Heidelberg, in the laboratories of both Kirchhoff and Bunsen, is in *The life & experiences of Sir Henry Enfield Roscoe* (ref. 4), 44–62 and 68–69.
8. Schuster, "Biographical byways"(ref. 7), 126, and *Biographical fragments* (ref. 4), 56, where the research laboratory is described as occupying a single room.

tion that colored the writings of contemporaries (among them educational reformers in Britain and France),[9] and it still directs much historical inquiry. But it is far from clear how accurately contemporary comments convey the realities of teaching and laboratory work in physics before about 1880. Recent studies have shown that even in Germany the institutionalization of research in the discipline and a systematic provision for the training of students in original work came about slowly and with considerable difficulty. In fact, the years between 1840 and the 1870s saw a constant struggle between physicists keen to improve the research facilities available to themselves and their students, and university councils and administrators who were reluctant to meet the cost involved.[10] Repeatedly, calls for laboratories and equipment on a scale and of a quality that professors were coming to regard as indispensable alarmed even their most sympathetic colleagues in other disciplines, who observed that the sums requested far exceeded what had been required to sustain the *mélange* of often old-fashioned apparatus that had made up traditional cabinets and served well enough for elementary teaching and demonstration, though nothing more.

Such a state of affairs is curious. For it might have been expected that in the years when research was assuming greater prominence in career-making and the quest for prestige, universities would be eager to place laboratories at the disposal of young physicists. Initially, however, this seldom happened. Indeed, in the 1840s and early 1850s, the most celebrated seminars and laboratories in physics still depended to a large extent on the private support they received from the chair-holder, and research remained essentially a personal activity, pursued and paid for by an individual and condoned rather than actively encouraged. In these circumstances, the occasional favored student might see the professor at close quarters working in his laboratory, but access to instruments would be limited to the rare occasions when he was honored by an invitation to collaborate.

One reason for this inconsistency in a system that prized research while failing to provide the facilities it required was that, however the criteria for

9. German influence on British debate about the reform of education and science is well treated in George Haines IV, *Essays on German influence upon English education and science* (Hamden, CT, 1969), esp. chapters 1–4. On French perceptions of Germany, see Harry W. Paul, *The sorcerer's apprentice. The French scientist's image of German science 1840–1919* (Gainesville, FL, 1972), esp. 1–14, and Robert Fox, "The view over the Rhine: perceptions of German science and technology in France, 1860–1914," in Yves Cohen and Klaus Manfrass, eds., *Frankreich und Deutschland. Forschung, Technologie und industrielle Entwicklung im 19. und 20. Jahrhundert* (Munich, 1990), 14–24, esp. 14–15.

10. David Cahan, "The institutional revolution in German physics, 1865–1914," *Historical studies in the physical sciences, 15:2* (1985), 1–65, on 3–12.

appointments might have changed by the 1850s, teaching remained the main duty of a professor, once he had taken up his post. Professors also had to reconcile themselves to the fact that, in almost any physics class, a high proportion of the pupils would be going on to careers not in higher education but in a secondary school, either a classical *Gymnasium* or one of the more scientifically oriented (and somewhat less prestigious) *Realschulen.* University boards, in turn, were therefore more disposed to provide funds for activities appropriate to the fashioning of future teachers than for research: their clear preference was for university curricula that promised the traditional Humboldtian virtues of breadth and exposure to a variety of sciences rather than the specialization that might have done more to advance the work of a young creative scientist.

So it was despite, and not because of, the circumstances prevailing in the university world that the first attempts to provide for laboratory research in physics were launched. A notable pioneer whose early difficulties illustrate the point was Wilhelm Weber. When in 1831, the young Weber was appointed to the chair of physics at the University of Göttingen, he requested equipment that would allow him to teach experimental technique and thereby to prepare the ablest students to go on to research. Despite having the influential backing of his very senior Göttingen colleague Karl Friedrich Gauss, Weber's case fell on deaf ears. The weakness of his case lay in his attempt to justify expenditure on instruments and laboratory space by invoking their ultimate importance for research, and it was only when he changed his strategy, to stress the needs of teaching, that he began to receive the funds he sought. Even then, however, success came slowly, and never abundantly. By the 1850s, during the period of his second appointment at Göttingen (following his earlier dismissal from his chair on political grounds), he managed to secure almost £50 a year for his laboratory and preparation and teaching rooms, in addition to occasional special allocations. But the sums he amassed were meagre and they left the facilities at Göttingen in the "somewhat primitive" state that Arthur Schuster noted when he spent two months there, just after Weber's retirement, in 1874.[11]

At the University of Königsberg another early pioneer in the cause of research, Franz Ernst Neumann, had similar experiences. From the time of his appointment as a *Privatdozent* (formally in mineralogy) in 1826, he made his ambitions for a serious commitment to research clear.[12] But when in 1834 he eventually succeeded in inaugurating his mathematical-physical seminar, with its accumulated apparatus (but no properly constituted physi-

11. Schuster, *Biographical fragments* (ref. 4), 64.
12. Kathryn M. Olesko, *Physics as a calling. Discipline and practice in the Königsberg seminar for physics* (Ithaca, NY and London, 1991), esp. 172–204.

cal laboratory), he found it difficult to establish its purpose as anything other than broad, relatively elementary instruction. By these older criteria the seminar was a success: between 1834 and his retirement in 1876 it trained no fewer than 150 students, roughly half of whom became teachers for all or a substantial part of their careers, mainly in the *Gymnasien*.[13] In contrast, however, Neumann's hopes of simultaneously preparing an élite of students for advanced research were thwarted by the unresponsiveness of university and governmental authorities to two decades of his relentless campaigning for a location and equipment. Finally, in frustration, Neumann resorted to the well-tried academic expedient of dipping into his own pocket. In this way, he managed to launch research in a house paid for from a legacy received by his second wife.[14] But the provision was far from extravagant, and as the number of students grew, the shortage of space and apparatus forced on him a renewed emphasis on pedagogy. It also obliged him to recognize that the main function of his seminar was, after all, to provide mainly theoretical instruction, complemented by only rather simple laboratory equipment and straightforward exercises in measurement.[15] Despite his frustration, Neumann's standards never wavered, and his teaching of experimental technique and the analysis of data remained in many ways a model until his retirement in 1876. But in the last quarter of a century of his career, the intellectual content of his seminar changed little.

Only in Berlin was the situation about the mid-century significantly more favorable to experimental research. However, the reason for this did not lie in the mechanisms of the University of Berlin, despite the fact that the university was conceived in 1809 with a view to its becoming a leading center for research and scholarship. It lay rather in the contingent circumstance of the appointment of the energetic Gustav Magnus as extraordinary professor in 1834 and then as ordinary (i.e., full) professor in 1845. Although Magnus managed to secure a regular allocation from the university for the maintenance and replenishment of the philosophical faculty's collection of physical apparatus, it was his private fortune which, for over two decades from the mid-1840s, allowed physics at the university to overcome the constraints of being limited to two modest rooms.[16] Throughout

13. Ibid., 317–365.
14. Ibid., 205–207.
15. Ibid., 205–313.
16. On physics in Berlin until the inauguration of the new institute in 1878, see Max Lenz, *Geschichte des Königlichen Friedrich-Wilhelms-Universität zu Berlin* (4 vols., Halle, 1910–1918) 3, 278–286. Magnus's achievements are reviewed in detail in the collective volume, Dieter Hoffmann, ed., *Gustav Magnus und sein Haus* (Stuttgart, 1995); see esp. Horst Kant, "Gustav Magnus und seine Berliner Physiker-Schule," 33–53.

this period, in which Hermann von Helmholtz, Gustav Wiedemann, August Kundt, Gustav Karsten, and John Tyndall were among the many physicists who worked at Berlin, research, advanced training, and seminar discussion all took place principally in Magnus's private laboratory (see fig. 3). Even when in 1867 Magnus persuaded the Minister to advance funds for a formally constituted institute, the premises remained in his house, and much of the work done there continued to depend on his personal contribution of facilities and money. It was only in 1878 that the Berlin physics institute, with Helmholtz at its head, could finally move into its new building from the temporary quarters in the university where it had been housed since Magnus's death in 1870. Admittedly, the building was spectacularly grand and well equipped, the like of which no other country in the world could boast. But, at the time, it was still very much the exception within Germany, only the relatively modest laboratory at Leipzig having preceded it as a purpose-built institute.[17]

Although most German physicists in the mid-nineteenth century had to plead pedagogical need in order to secure the resources, other than those they could supply themselves, for their own and their pupils' research, it would be wrong to conclude that the importance of teaching was invoked simply for rhetorical effect and in pursuit of quite different aims. In Germany as elsewhere, the conviction that laboratory exercises had an essential role in teaching was genuine: most teachers of physics did not doubt that a direct, and directed, engagement with the laws of their discipline provided the only route to a full understanding of physical principles. Professors recognized, however, that laboratory exercises, in particular in precise measurement, not only served the purposes of basic instruction but also prepared the ablest pupils for the more advanced manipulations on which any professional career in physics would have to be based. Hence, in practice, the boundary between the use of the laboratory for teaching and for research could easily be blurred, and it frequently and quite naturally was when professors engaged in their pursuit of funding.

In this pursuit, rivalry between universities made a constructive contribution. As early as the 1850s, in the competition to attract the most distinguished physicists, the offer of a well-equipped *Praktikum* (usually a laboratory, a teaching room, and space for preparation and storage) served as a potent lure. But the improvements that occurred then and for some years afterwards were modest, and it was not until the heady days following Prussia's victory over France in 1870 and the consolidation of the new Reich that the claims of physicists, along with those of other academic scientists, began to bear the kind of fruit that were to become the envy of other nations in the 1880s and 1890s.

17. Cahan, "The institutional revolution" (ref. 10), 12–25.

It was only in this later phase, as we argue in the last section of this chapter, that the justification for the escalation of expenditure significantly changed its character, and that cases founded on the interests of pedagogy lost the weight they had once had. The change owed much to a weakening of the demand in secondary education for the products of the university seminars. No less importantly, however, it also reflected the fact that the champions of university science in general and physics in particular were finding a more potent justification for their disciplinary ambitions in what they could now purvey as the scientific needs of industry. In this rhetorical shift, which privileged a new form of utility, physics departments in Germany were not unlike those in other nations. The great difference lay, as we shall see in the rest of this chapter, in the far greater benefit that they drew from the change in strategy.

France: Science, bureaucracy, and reform

Although laboratory-based research had its most impressive results in the German universities in the later nineteenth century, it was in France that research degrees first emerged as an ingredient in academic career-making. The *doctorat-ès-sciences* and the *doctorat-ès-lettres* were both created in 1808 as part of the administrative structure of Napoleon I's new Université Impériale, and thereafter they were designated as necessary qualifications for the holders of chairs in the faculties of science and letters. Despite the clarity of the regulations, however, the letter of the law proved hard to implement. For the first twenty years or so only a small minority of professors in the sciences and mathematics held a doctorate, and the degree did not become a common element in the *cursus* leading to a chair until the 1830s. Even then, implementation of the formal requirement was imperfect, and it remained so until the 1860s, the first decade in which all of the new professors appointed to the faculties of science held the degree.[18]

Since the doctorate required the performance of independent work and the writing of a printed thesis, it served to focus the attention of aspirants to chairs on research and publication. Before the Second Empire, however, most theses in science and mathematics were perfunctory affairs that called for precise but simple measurement or the careful treatment of an uncontentious problem, with significant new departures in either experimental technique or theory remaining rare exceptions. Theses also tended to be short. Sixty-three per cent of them before 1850 were no more than

18. Robert Fox, "Science, the university, and the state in nineteenth-century France," in Gerald L. Geison, ed., *Professions and the French state, 1700-1900* (Philadelphia, 1984), 66-145, on 76.

FIG 3. Gustav Magnus's house in Berlin. Until 1870, the first physical laboratory in Germany was housed in this building. Reproduced with permission of the Hochschule der Künste Berlin, Hochschularchiv.

forty pages long.[19] It is a mark of the increasing seriousness of theses that this proportion fell to 16 per cent between 1850 and 1869, and to 7 per cent after 1870, and that in the same period the average age of *thèsards* rose appreciably. Yet the fact remains that an essentially routine exercise, efficiently performed, could still win a doctorate as late as the 1870s.

In these circumstances, the facilities available to candidates caused no serious concern, and most aspirants to the doctorate conducted their research with apparatus of a kind that could be found or improvised in a run-of-the-mill school laboratory. Here, even the most distinguished academic careers could be launched, as the case of the physicist Jules Jamin demonstrates. His notable doctoral thesis of 1847 on the reflexion of light from metallic surfaces grew from research he performed while serving as a teacher in secondary education first at the Collège Royal at Caen and then

19. Victor Karady, "Educational qualifications and university careers in science in nineteenth-century France," in Robert Fox and George Weisz, eds., *The organization of science and technology in France 1808–1914* (Cambridge and Paris, 1980), 95–124, on 111n.

at the Collège Bourbon and the Collège Louis-le-Grand in Paris.[20] The immediate aim of those who pursued research in these circumstances was almost invariably to pass a bureaucratic hurdle and obtain a qualification that would ease the passage from the classroom to an administrative post or (as in Jamin's case) to a chair in higher education.[21] There was no suggestion that doctoral research should respond, except in the most general way, to the shifting frontiers of the discipline, still less to the interests of the economy.

The academic reform movement that swept France in the 1850s and, more particularly, the 1860s and early 1870s created the first concerted resentment against the neglect of original research and scholarship and the indifference towards the conditions available for such work.[22] The movement had a broad base. But its main roots and support lay in the national university, the Université de France, in particular in the faculties of science, letters, and medicine, and among those of a politically liberal persuasion who opposed what they portrayed as the superficiality of French intellectual life during the Second Empire. In these circles, both within and outside the Université, there existed a strong admiration for German academic life and the critical spirit that it was seen to engender. For the humanities, the sentiment was articulated with incomparable conviction and vehemence by Ernest Renan, the professor of Hebrew at the Collège de France, who was expelled from his chair in 1864 in the wake of the violent Catholic reaction to his *Vie de Jésus*, a book that scandalized bourgeois and clerical opinion by presenting Christ as a secular figure.[23] It was in the sciences, however, that the vision evoked its keenest response. Here, one natural champion was Renan's close friend, Marcellin Berthelot, professor of organic chemistry at the Ecole de Pharmacie in Paris (from 1859) and then (from 1865) at the Collège de France. But Berthelot's mixture of respect and envy for the intellectual achievements of the German universities was widely shared, especially in the generation of *savants* who reached the peak of their academic careers in the later years of the Empire. Of this genera-

20. "Mémoire sur la réflexion métallique," in Jules-Célestin Jamin, *Thèses de physique et de chimie présentées à la Faculté des Sciences de Paris* (Paris, 1847), 5–53.
21. Jamin was appointed professor of physics at the Ecole Polytechnique in 1852 and at the Sorbonne in 1863.
22. Fox, "Science, the university and the state" (ref. 18), 85–101; George Weisz, "Le corps professoral de l'enseignement supérieur et l'idéologie de la réforme universitaire en France, 1860–1885," *Revue française de sociologie, 18* (1977), 201–232, on 227–229, and *The emergence of modern universities in France, 1863–1914* (Princeton, NJ, 1983), 59–89 and 196–224.
23. Claude Digeon, *La crise allemande de la pensée française (1870–1914)* (Paris, 1959), esp. 179–215. Digeon's book is a rich source on the place of German thought in the reform movement in France before as well as after 1870.

tion, no one was more authoritative in his espousal of the cause of science than the Alsatian chemist Adolphe Wurtz.

Wurtz's case, in the form of an official report on the facilities of the German universities for research and teaching in chemistry, physiology, and anatomy, completed in 1869 and published in the following year, rested on accounts of more than a dozen laboratories in some of the leading German universities.[24] Wurtz used the accounts to convey an impression of a level of material support and a commitment to research, on the part of governments and individual scientists, that had no counterpart in France.[25] Since the mid-century, Wurtz argued, German scientists had benefited both from values in university and public life that had come to favor advanced study and originality, and, at least in chemistry and medicine, from the facilities that were necessary to turn those values into reality. No indictment of the neglect of laboratory provision in France could have been more direct or telling. Amid the heightened sensitivities provoked by the defeat of 1870, when the chemist Henri Sainte-Claire Deville, the mathematician Charles Hermite, the anthropologist Quatrefages de Bréau, and the chemist Jean-Baptiste Dumas all addressed the Académie des Sciences with one highly critical voice,[26] it gave dismal substance to the more subjective account of the "suffering" and "penury" of the French scientific community of which Louis Pasteur had already spoken. The fault, as Pasteur had displayed it in a devastatingly critical pamphlet in 1868, lay in the low priority that experimental science had been given in France for over half a century, since the golden age of the First Empire.[27] For Pasteur, as for Wurtz, Germany had become the pacemaker, and France the laggard whose damp, unhealthy laboratories of the kind that were said to have hastened the death of Armand Bineau, a professor of chemistry in Lyon, were the symbol of the country's decline.[28]

The confidence with which Pasteur and Wurtz put their case owed much to the encouragement they received from an influential ally, Victor Duruy. As a liberally minded Minister of Public Instruction from 1863 to 1869, a strong though clear-eyed champion of the faculties, and an ancient histo-

24. Charles-Adolphe Wurtz, *Les hautes études pratiques dans les universités allemandes. Rapport présenté à Son Exc. M. le Ministre de l'Instruction Publique* (Paris, 1870).
25. Ibid., 11–13. Wurtz pursued his inquiries after 1870, publishing a second report, on nine more laboratories, which appeared as *Les hautes études pratiques dans les universités d'Allemagne et d'Autriche-Hongrie. Deuxième rapport présenté à M. le Ministre de l'Instruction Publique* (Paris, 1882). Here, see esp. his letter to the Minister of Public Instruction on pp. 1–33.
26. *Comptes rendus hebdomadaires des séances de l'Académie des Sciences, 72* (1871), 237–239 and 261–269.
27. Louis Pasteur, *Le budget de la science* (Paris, 1868).
28. Ibid., 7.

rian versed in the practice of scholarship, Duruy exerted discreet but firm pressure in support of research.[29] Although the cautious conformity of Napoleon III's glittering but shallow Empire limited his room for maneuver, it was he who urged Pasteur to write his pamphlet of 1868 and who in the same year commissioned Wurtz's report on the German laboratories. Moreover, by establishing the Ecole Pratique des Hautes Etudes, devoted to research across the whole range of the scientific and humanistic disciplines, he achieved in 1868 an enduring triumph that bore rich fruit long after his departure from office.

Although the Ecole was a purely administrative structure that functioned through the activities it was able to stimulate in existing laboratories, it did much to give scientific research greater visibility.[30] In its early years, a chronic shortage of funds gravely impeded its development. But from the 1870s, the Ecole served to divert significant, if never large sums for the support of advanced work. Its original, predominantly cultural aims received at that time crucial reinforcement, across the range of the scientific disciplines, from two new sources. One was the widespread conviction that the advancement of science would serve to rehabilitate national morale following the humiliation of 1870 at the hands of the Prussians. The other, less precisely articulated but reflecting a new perception emerging throughout the industrial world, was the conviction that science should be supported for the benefit it yielded for the economy.

Physics emerged as a leading, and arguably the greatest, beneficiary of this re-evaluation of the role of academic research. From the start, two physics laboratories, both housed in the cramped quarters of the old Sorbonne, formed part of the Ecole's federal structure. One, the Laboratoire des Recherches Physiques, was under the supervision of Jamin, who held a chair of physics not only at the Sorbonne but also at the Ecole Polytechnique; the second was directed by the other professor of physics at the Sorbonne, Paul Desains. In both laboratories, the creation of openings for assistants allowed at least a few young men to avoid the traditional rite of passage as a teacher in secondary education and to devote the early stages of their academic careers more fully to research. Among those who benefited were Edmond Bouty and Gabriel Lippmann, who worked under Jamin, and (during his brief stay as a young *chef de travaux* and then

29. Jean Rohr, *Victor Duruy, ministre de Napoléon III. Essai sur la politique de l'instruction publique au temps de l'Empire libéral* (Paris, 1967), esp. 116–123, and Robert Fox, "Science, the university, and the state" (ref. 18), 96–101.

30. On the Ecole and Duruy's role in its establishment, see Harry W. Paul, *From knowledge to power. The rise of the science empire in France, 1860–1939* (Cambridge, 1985), 44–50. Most of the detailed information that follows is taken from the annual *Rapport sur l'Ecole Pratique des Hautes Etudes*.

directeur adjoint under Desains) the future contributor to wireless telegraphy Edouard Branly. Although the environment was never opulent, it helped to foster both individual reputations and a keen sense of team-work. Branly, in particular, was stimulated by Desains's personal example and through him came into regular contact with a peer group of other young physicists, including some, such as Jules Violle and Emile Amagat, who went on, like Branly, to distinguished academic careers.

The style of work pursued in the two laboratories differed considerably. But both of them reflected the growing importance that, in France as elsewhere, was coming to be attached to experiment in the practice and pedagogy of physics. Jamin devoted his laboratory exclusively to research at the highest level, admitting only pupils, normally working for a doctorate, who were aspiring to academic careers. The numbers working under Jamin were never large—typically far fewer than a dozen in the 1870s—but foreign observers, including George Carey Foster and Arthur Schuster from Britain, as well as contemporaries in France, regarded it as one of the leading European centers of research in physics. By contrast, the success of Desains's laboratory, the Laboratoire d'Enseignement de Physique, was more directly measured by the number of the students he attracted: by the mid-1870s, he had over fifty students working with him, and although the laboratory had twenty or more rooms, conditions were cramped. He emphasized teaching and the supervised conduct of routine exercises by candidates preparing for the first higher degree, the *licence,* and the competitive examination for teachers in the *lycées,* the *agrégation.*[31] In both laboratories the range of interests was wide, though with some emphasis on electricity and magnetism and optics in Jamin's group and on radiant heat, latent heat, and spectroscopy and other areas of optics in Desains's. Significantly and predictably, there was a shared orientation towards precise measurement, chiefly of constants and other data required in the verification of physical laws.

Other Parisian laboratories also developed strongly between the 1860s and 1880s. In physics, at the Muséum d'Histoire Naturelle, the laboratory of Antoine-César Becquerel and his son Edmond built on its attachment to the Ecole Pratique des Hautes Etudes, creating a strong reputation for electrical research from the early 1870s. Chemistry too benefited. Henri Sainte-Claire Deville's laboratory at the Sorbonne and Pasteur's at the Ecole Normale Supérieure were just two of several in the discipline that attracted

31. An excellent account of the training of teachers for careers in the *lycées,* with special reference to the *agrégation,* is in Nicole Hulin-Jung, *L'organisation de l'enseignement des sciences* (Paris, 1989), 191–221.

students and acquired a reputation for originality, with the aid of the support that came to them from the Ministry of Public Instruction as constituents of the Ecole. The decade following the war of 1870 proved to be a particularly fertile one for the consolidation of what had been achieved at the very end of Duruy's reign and for some expansion. As a result, by the late 1870s, the Ecole could boast almost forty scientific laboratories (the great majority of them in Paris) receiving support across the whole range of the physical sciences, geology, the life sciences, and medicine.[32] In addition, all fifteen of the country's faculties of science had laboratories for the main scientific disciplines, as did the Ecole Polytechnique and the other technical *grandes écoles*.

This broad diffusion of facilities had an undoubted effect on attitudes towards research and on the vigor with which research was pursued: the marked growth in the number of candidates for the *doctorat-ès-sciences* from the mid-1870s, after a quarter of a century of relative stability, is one indicator of this, as is the simultaneous increase in the number of publications emanating from the nation's faculties of science.[33] Yet, until well into the 1880s, and arguably beyond, the physical conditions remained modest. In fact, when the eminent educational administrator Octave Gréard looked back in later life on facilities at the "old" Sorbonne, they appeared, in retrospect, to have been almost spectacularly primitive. Of a dismally memorable visit on which Desains had been his guide in the 1870s Gréard wrote:[34]

> One day... during a visit, with the professor of physics Paul Desains, to the rooms, with their low, sloping roofs, that had been allocated to him—as

32. A useful summary of the decisions integrating laboratories in the Ecole Pratique des Hautes Etudes between 1868 and 1877 is in *Statistique de l'enseignement supérieur. Enseignement, examens, grades, recettes et dépenses, en 1876. Actes administratifs jusqu'en août 1878* (Paris, 1878), 707–728.
33. The number of doctorates awarded by the Faculté des Sciences in Paris rose to 15 a year for the five years 1875–1879, having remained quite stable at about 8 since 1850. The figure rose to 18 between 1880 and 1884 and over 30 between 1885 and 1889. See Albert Maire, *Catalogue des thèses de sciences soutenues en France de 1810 à 1890 inclusivement* (Paris, 1892), 38–129. Maire's list for the provincial faculties shows numbers there to have been very much lower. On the growth in the number of publications by professors in the faculties, see below, pp. 130-133.
34. Octave Gréard, "Introduction," in H.-P. Nénot, *Monographie de la nouvelle Sorbonne* (Paris, 1903), 19–20. Cf. the similar sense of excitement conveyed in Gréard, *Nos adieux à la vieille Sorbonne* (Paris, 1893), 245. For further evidence of the cramped conditions, however, see *Ecole Pratique des Hautes Etudes. Rapports des directeurs de laboratoires et de conférences 1865–1877* (Paris, 1879), 19–21.

> he showed me the windows that would not close properly, the ill-fitting doors, the sweating red tiles on the floor, the worm-eaten stairs on which at almost every step he had to interrupt his explanations with a watchful "Be careful!"—he stopped in a dark corner and said: "Despite all this, how good it is to be here."

In Gréard's recollection, the squalor he described had gone hand in hand with a feeling of achievement. The dilapidated attics and sheds ("hangars") of the cour Saint-Jacques had been the "cradle of the renaissance," where Jamin had developed his "brûleur électrique" for Paul Jablochkoff's highly successful arc lamp and Gabriel Lippmann had taken some of the first steps in color photography.[35] Above all, they were places where the promotion of experimental technique and a spirit of collective endeavor had flourished, regardless of the makeshift material environment.

Even when Gréard wrote (in 1903, in the introduction to Philippe Nénot's magnificent description of the new buildings that now graced the Sorbonne), there were still those who maintained, with justice, that the French educational system lagged significantly behind Germany in the material provision it made for science.[36] But the changes that had come about since the modest first steps taken at the end of the Second Empire were still profound. Encouraged, from the start, by the administrations of the Third Republic, and reinforced, especially after 1880, by the injection of private funding and grants from municipal and departmental authorities, they had resulted in a rebuilding program that by 1900 had transformed the faculties, in particular the faculties of science, throughout France. In the process, research and laboratory teaching had been confirmed, at least in principle, as fully institutionalized elements of a healthy academic life and hence as activities that had a proper call on funds. Pierre Duhem was one critic at the turn of the century who believed that the expenditure had been largely misdirected to decoration and show: citing the case of his younger colleague in the Faculty of Science at Bordeaux, Albert Turpain, who had been obliged to pursue his doctoral research on electromagnetic waves in space generously provided by a local industrialist, he excoriated a movement that had provided the university system with "monuments"—

35. Gréard, "Introduction" (ref. 34), 20.

36. For a helpful comparative study of expenditure on the laboratories and support facilities for physics in the main scientific nations about 1900, see Paul Forman, J. L. Heilbron, and Spencer Weart, "Physics *circa* 1900: personnel, funding, and productivity of the academic establishments," *Historical studies in the physical sciences*, 5 (1975), 1–185, esp. (on France) 4–10, 66–68, and 125–127.

"a palace today, a constraint tomorrow, a tomb the day after"—rather than flexible working spaces.[37] Others argued that the ever stronger bond between experimental physics and the interests of industrial utility (to which Duhem gave a qualified welcome) represented a threat to the discipline.[38] But whatever strategic errors might have been made along the way, the fact remains that, from unpromising beginnings in the 1860s and through a convergence of professional aspirations, economic circumstances, and the desire to reassert the intellectual dignity of the nation, science in general had become far stronger as a field for research in France. Moreover, and crucially for our argument, experimental physics had moved to the center of the pantheon of the sciences.

Britain: Education and research in the age of inquiries

In the mid-nineteenth century Britain stood firmly among the leading scientific nations, with substantial achievements in research, especially in physics and mathematics, to her credit. There was, nevertheless, a persistent and widely felt sense of decline. This focussed primarily on the way in which the more experimental aspects of research were organized. In a manner reminiscent of the situation criticized by Charles Babbage in 1830, voluntarism and self-help were still the bed-rocks of British scientific achievement.[39] By and large, individual researchers were expected to display resourcefulness in their search for ways of sustaining their work, ranging from the gentleman-of-leisure style exemplified in the research of James Joule to projects sponsored by large railway companies and other industrial interests.

The dominance of this characteristically British style does not mean that the central authorities were wholly unresponsive to the needs of science.[40] In Britain as elsewhere, a number of posts were sustained directly by gov-

37. Pierre Duhem, *Usines et laboratoires* (Bordeaux, 1899), 2–3. The text of this pamphlet had originally appeared in the *Revue philomathique de Bordeaux et du Sud-Quest,* 3e année, no. 9, 1 Sep 1899.
38. See, for example, the very critical lament by a champion of traditional academic values, Emile Picard, in his "La science et la recherche scientifique," *Revue scientifique,* 50e année, 2me semestre (1912), 577–581.
39. Charles Babbage, *Reflections on the decline of science in England, and on some of its causes* (London, 1830). On these characteristics of British science in 1830, see J.B. Morrell, "Individualism and the structure of British science in 1830," *Historical studies in the physical sciences,* 3 (1971), 183–204.
40. An excellent general account of the patronage of science in Victorian Britain is William H. Brock, "The spectrum of science patronage," in Gerard L'E. Turner, ed., *The patronage of science in the nineteenth century* (Leiden, 1976), 173–206.

ernmental funding, notably in the Geological Survey and at the Greenwich Observatory. And awards, some of them substantial, were made for projects that were deemed to be prestigious or of public utility. In this process, the British Association for the Advancement of Science was especially important. From its foundation in 1831, it acted not only as a source of minor grants from its own resources but also as a gadfly for the elicitation of *ad hoc* governmental awards: William Whewell (for his work on tides) and the Reverend Henry Moseley (for his study of the stability of ships) were among those who benefited from this rather untidy mechanism in the 1830s and 1840s.[41] From 1849, the government also contributed to research through a grant of £1000 per year (administered by the Royal Society) intended as support for the work of individual scientists.[42] But even when this sum was augmented by a new annual parliamentary grant-in-aid of £4000 in 1876, the provision for research remained modest, and although the scheme continued after an initial trial period of five years, the individual amounts received by the growing number of applicants assumed a mainly symbolic character, of benefit only to small-scale inquiries.[43] In purely academic terms, the results, measured by the number of publications, were not negligible. Yet they left the need for a coordinated system for promoting scientific research at the national level unsatisfied.

The most obvious weakness lay in the absence of a recognized institutional setting. For while many of those who were most active in research held positions in higher education, their institutions provided little by way of incentives or material support. In the ancient universities of Oxford and Cambridge, degrees in science (as opposed to mathematics) only made their appearance in the 1850s and for some time attracted few candidates; even more importantly, these universities remained wedded to the traditional priorities of teaching and the refinement of learning rather than the pursuit of new knowledge.[44] The Scottish universities and the London-based colleges were more receptive to the growing importance of scientific education, but even this did not entail a recognition of the advancement of

41. The work of Whewell and Moseley was published in the *Philosophical transactions of the Royal Society of London* between 1833 and 1850.

42. Roy MacLeod, "The Royal Society and the government grant: notes on the administration of scientific research, 1849–1914," *Historical journal, 14* (1971), 323–358, on 325–345, and "Resources of science in Victorian England: the endowment of science movement, 1868–1900," in Peter Mathias, ed., *Science and society 1600–1900* (Cambridge, 1972), 111–166, on 113–115.

43. MacLeod, "The Royal Society and the government grant" (ref. 42), 337, and "Resources of science in Victorian England" (ref. 42), 140–146.

44. Donald S. L. Cardwell, *The organisation of science in England* (2nd edn., London, 1972), 85–98.

science as one of the duties of the professoriate. Here, too, teaching was an overriding preoccupation, not least because both the Scottish universities and the London colleges largely depended, for their income, on student fees.

The marginal place of research recurred frequently in public debate as the traditional styles of British higher education underwent a new wave of criticism from the late 1860s in what Donald Cardwell has characterized as "the age of inquiries."[45] The members of the Devonshire Commission (1870–1876) and the Samuelson Commission (1881–1884), in particular, took as a main target both the uncritical and repetitive approach to learning that was fostered, at all levels, by the structure of examinations, and what they identified as a damaging separation of teaching from research.[46] This view rested on the same belief in the pedagogical benefit of instruction in laboratory techniques, notably though not exclusively in precise measurement, that characterized attitudes to the physical sciences in Germany and France. And, as in those countries, it lent welcome support to the attempts of professors to secure the ever costlier equipment and assistance they needed for the cultivation of their own research. However, in the absence of governmental support, it proved extremely difficult to bridge the gap that academic physicists identified between the inadequate existing provision and what they portrayed to their paymasters and the lay public as desirable.

As the case of chemistry had already indicated, a tempting solution lay in a liaison with industry and with individual manufacturers and local authorities keen to promote economic growth. Here, chemists had long been aware that some of the most fertile areas of research lay precisely at the interface between science and industrial practice, and during the last quarter of the century, any doubts that physics should go down the same road were gradually (though never completely) eroded. Predictably, the *rapprochement* between research and industry provoked suspicion, sometimes amounting to overt hostility, among dons of the old school, jealous of the remote, intellectualist traditions of their calling.[47] Others, however, eagerly supported the movement, some because they found it genuinely congenial, some for more pragmatic reasons reflecting the harsh reality that a rhetoric based on the value of research for its own sake would never elicit the necessary support, whether from the institutions of higher education, from private patrons, or from government. Whatever their motives, such physicists turned to the areas of their disciplines in which experimen-

45. Ibid., 111–155.

46. Ibid., and MacLeod, "Resources of science in Victorian England" (ref. 42), 126–139.

47. See the comments of Emile Picard, cited in note 38.

tal techniques were most easily adaptable to the realm of practice and in which laboratories purveying those techniques could present themselves as offering valuable expertise and a relevant preparation for industrial as well as academic careers. In this way, electricity emerged as a particularly favored field, one that combined the promise of utility with intellectual excitement, and it is no coincidence that Britain resembled all other industrialized countries in the leading role that this branch of physics played in the development of university- and college-based laboratories.

In this phase of development, departments of chemistry not only served as a model for what now began to occur in physics; they also provided a seedbed for the new generation of men who, in the 1870s and 1880s and across the whole range of the sciences, led the campaign for the institutionalization of laboratory instruction. The chemists' style of work, with its emphasis on the introduction of students to the value of precision and good bench technique, was carried over, to good effect, into the practices of George Carey Foster (at University College, London), Frederick Guthrie (lecturer in physics at the Royal School of Mines in London), and others who moved into physics after an initial training in chemistry.[48] An alternative influential model was that of William Thomson, then in his prime as professor of natural philosophy at the University of Glasgow. Thomson's achievement, however, was difficult to emulate in every detail: his particular personal charisma and public standing were essential to the success of the Glasgow laboratory, as was the support that he channelled into it through his connections with telegraphic and cable companies. Where the circumstances were immediately less favorable, laboratories had to be built up in different ways, though always with an eye on the problem of garnering resources in the face of multiple competing claims. The provision of relatively low-level teaching was the commonest answer to the challenge. In this respect, Foster's strategy as professor of experimental physics at University College, directed at the accumulation of a substantial income from student fees, was typical. Soon after his appointment to the chair in 1865, he set up a small laboratory in which he offered a course focussed primarily on the techniques of precision measurement. Here, in parallel with his practical classes, he privately pursued his work for the BAAS Committee on electrical standards and from time to time engaged some of his

48. Foster's comment in his obituary of Guthrie makes the point explicitly. He saw Guthrie as typical of those "physicists who had served their time as chemists"—largely by necessity at a time (before the 1860s) when the chemical laboratory had been the only available setting for a training in experimental practice; see *Nature*, *35* (1886–1887), 8–10. On Foster's career, see the obituary of him by "O.J.L." [Oliver Lodge] in Royal Society of London, *Proceedings*, *96* (1919–1920), "Obituary notices of Fellows deceased," xv–xviii.

best students to assist him, though always maintaining teaching as the laboratory's main activity.

A similar need to secure running costs once the initial investment had been made colored the history of most early physics laboratories. At Oxford, the professor of experimental philosophy, Robert Clifton, started by securing the handsome sum of £10,700 from the trustees of the eighteenth-century bequest of the great-grandson of the Earl of Clarendon, in order to construct a new laboratory—the first purpose-built physics laboratory in the world—that opened in 1870.[49] Clifton aimed to implant in Oxford a tradition of both teaching and research in experimental physics that would have set the Clarendon Laboratory ahead of most of its rivals. But his auspicious beginning yielded disappointing results. As further support for apparatus, additional professors, and assistants failed to materialize, Clifton's aspirations were steadily reduced and finally, in 1887, dramatically curtailed, when the University refused his request for £4,800 with which to build what he rightly regarded as a much-needed electrical laboratory.[50] Thereafter, research succumbed inexorably to the overwhelming demands of elementary teaching, until Clifton rather theatrically announced in 1897 that he had been obliged to reallocate his personal laboratory for undergraduate use as an electrical laboratory and that he saw no alternative to either finding new premises for his own equipment (which he now removed from the laboratory) or withdrawing completely from original work.[51] A year later, he made his decision, announcing that, in the absence of space in which to perform research, "all attempts to undertake work of this character must in future be abandoned."[52]

In telling contrast with Oxford, the Cavendish Laboratory in Cambridge successfully maintained the delicate balance between academic respectability, the kind of utilitarian activities that would win visibility and in-

49. On the Clarendon Laboratory and the work that was conducted in it, see Graeme J.N. Gooday, "Precision measurement and the genesis of physics teaching laboratories in Victorian Britain" (University of Kent Ph.D. thesis, 1989), chapter 6, and Robert Fox, "The University Museum and Oxford science, 1850–1880," in Michael G. Brock and Mark C. Curthoys, eds., *The history of the University of Oxford*. Vol. 6. *Nineteenth-century Oxford, part I* (Oxford, 1997), 641–691, on 676–682. The total cost of the laboratory, including the initial expenditure on apparatus, was about £12,000; see Fox, 682n.
50. Janet Howarth, "Science education in late-Victorian Oxford: a curious case of failure?," *English historical review*, *102* (1987), 334–371, on 343.
51. *Oxford University gazette*, 23 Jun 1897, 632.
52. *Oxford University gazette*, 7 Jun 1898, 588.

come, and the promotion of research.[53] Completed in 1874, at a cost of £8450 and with the generous support of Sir William Cavendish, seventh Duke of Devonshire and Chancellor of the University of Cambridge, the laboratory benefited from the engagement of its first director and the first holder of the University's chair of experimental physics, James Clerk Maxwell, in the determination of absolute units of electrical resistance.[54] From its start under the aegis of the BAAS in the early 1860s, this work had involved physicists in other universities, not only Foster in London and Thomson in Glasgow but also Balfour Stewart in Manchester, following his appointment to the chair of physics at Owens College in 1870. When the Cavendish opened, however, Maxwell achieved an immediate coup by making the laboratory the main national center for the project. Partly through the high objectives that he set, he secured acceptance for the laboratory in conservative Cambridge circles, and soon the reputation of the Cavendish spread. Following his death in 1879, his successor, Lord Rayleigh, continued to bring in instruments and talents of the highest quality and simultaneously to promote both advanced research and more humdrum work on units and standards. The latter work, which allowed students, assistants, and instrument-makers to engage in an activity equally vital as a tool for scientific inquiries and for industrial practice, was no less important than the former in the laboratory's rise to the status of Britain's leading center for research in physics at the end of the nineteenth century.

In the provincial university colleges, the achievements in physics could not match those of the Cavendish. But Stewart at Owens College was typical of the new generation of professors who, like his peers in Cambridge, saw the encouragement of research as a responsibility that should, and could, go hand in hand with teaching. Soon after his arrival, the governing body of Owens assigned some of the funds that had been accumulated for the extension of the college to the fitting out of a physics laboratory: although the earmarked sum was modest, it allowed a course of practical physics to be introduced in 1871. As in Oxford and Cambridge, the emphasis was on precise measurement, extending to a variety of fields, including optics, heat, and, inevitably, electricity. With an eye on research, Stewart also encouraged at least some students to apply the techniques

53. The main general histories of the Cavendish Laboratory are *History of the Cavendish Laboratory 1871–1910* (London, 1910), and James Gerald Crowther, *The Cavendish Laboratory 1874–1974* (London, 1974).

54. On Maxwell's interest and the Cavendish as a "metrological institution," see Simon Schaffer, "Late Victorian metrology and its instrumentation: a manufactory of Ohms," in Robert Bud and Susan E. Cozzens, eds., *Invisible connections. Instruments, institutions, and science* (Bellingham, WA, 1992), 23–56.

they had acquired in the laboratory course to more advanced exercises. According to his first report on the laboratory, for 1871–1872, eight students had regularly attended the basic course, among them three of particular note: Thomas S. Tait, J.R. Ashwell, and Arthur Schuster (who was eventually to succeed Stewart in the chair).[55] All three had received at Owens a solid grounding in the theoretical as well as the experimental aspects of the subject, and although they did not formally have the status of advanced students, Stewart judged them sufficiently trained to engage in original work of considerable delicacy. It is a telling indication of the new value that administrators attached to research that the work of Tait, Ashwell, and Schuster (especially Schuster's on the spectrum of nitrogen) had as much prominence in the report as Stewart's.[56]

Despite its quickening rhythm, research alone could not sustain the development of the material provision for physics. The dominant role of teaching and the heavy reliance of universities, departments, and individual professors on student fees meant that the expansion of a laboratory or the opening of a new one remained difficult unless some compelling external circumstance, in particular a change of syllabus, intervened to enhance the importance of experimental classes. For this reason, until the diversity of provision increased towards the end of the nineteenth century, when the provincial university colleges began to receive independent status, the policies of the University of London exerted a crucial influence. The university's B.Sc. degree stood at the pinnacle of the aspirations of the science students in the provincial colleges, from its introduction in 1858. Hence when support for a hands-on approach in science teaching was endorsed in the revised structure of the university's examinations in 1875, the effect on the colleges was considerable. Under the new regulations, candidates for the B.Sc. honors degree and the higher degree of Doctor of Science, normally attempted two years later, were examined not only in the theoretical aspects of science but also through a practical test. As a result, the network of institutions that prepared for the London degrees suddenly had no choice but to offer training in experimental technique.

55. "Report on the physical laboratory during the session 1871–2," in *The Owens College, Manchester. Calendar for the session 1872–3* (Manchester, 1872), 82–86.

56. Ibid., 84 and 86. It should be stressed that not all professors of physics shared Stewart's belief in the benefit of a close contiguity between routine instruction and the active engagement of advanced students in independent inquiries. For example, when William Grylls Adams fitted out a laboratory and opened it to students soon after his appointment to the chair of natural philosophy at King's College, London, in 1865, the teaching he offered, though highly successful, remained firmly rooted in the regular curriculum in experimental physics; emphasis was placed on the use of instruments, but the instruction never served as an initiation in research.

Although the resulting proliferation of laboratory courses enforced some improvement, the pressure on space and equipment soon grew to take up any slack there might have been. Hence time and funds that could be devoted to research still had to be carefully accumulated and protected. The position of research was also constrained by the fact that the main demand for teaching came either from undergraduates at a very elementary stage in their careers or from students not seeking degrees in physics, while advanced students in the subject still constituted a small minority of the total enrolments until the first world war. The balance did not unduly disturb university administrators, for whom introductory courses, in particular those frequented by medical students, were far more profitable than those for students specializing in physics. But it had the effect of keeping professors of physics and their assistants (as, for the same reasons, it also kept their colleagues in chemistry) further than they would have wished from the cutting edge of their discipline.

In the circumstances, the passage of a student into research was a difficult, often chancy affair. It depended almost entirely on the kind of *ad hoc* academic apprenticeship that Stewart administered at Owens College and on an untidy mechanism of personal invitations that gave the occasional favored pupil an *entrée* to his professor's laboratory, either as an assistant or as a demonstrator. Where such an invitation was not extended, admission to a laboratory—of chemistry as well as of experimental physics—was possible in principle. But regular attendance was costly, and scholarships were few and small. Such innovations as the special fellowships that Trinity College, Cambridge, offered from 1867 to young graduates who wished to pursue research, and the Platt Physiological Scholarships at Owens College, inaugurated in 1872, helped to maintain a trickle of students entering research. Nevertheless, the number of those who took this path remained small until research degrees were established, beginning in the 1880s.

The first of the research degrees was the University of London's new type of D.Sc., instituted in 1880. The purpose of the doctorate that had been awarded since 1860 was to ascertain that the candidate (who had to have passed the B.Sc. two years or more before) had "a thorough practical knowledge" of at least one "principal subject," chosen from sixteen specified sciences, together with a "general acquaintance" with a subsidiary subject or subjects specified as being related to the main science selected.[57] This was achieved by a written examination based on questions at a more advanced level than those set for the B.Sc. honors degree, and even when a

57. The regulations for the various degrees were summarized in the annual *Calendar* of the University of London.

practical examination was introduced in 1875, there was no attempt to test originality or skills in research. The research-based D.Sc. of 1880 therefore marked a fundamental new departure, bred of a discussion inaugurated three years earlier about the role of the University of London across the whole spectrum of higher education.[58] In this discussion, the fact that the university was exclusively a body for the administration of examinations (as it had been since its creation in 1836) and that it had no direct responsibility for teaching was perceived to be a weakness at a time when traditional examinations as a means of assessing academic achievement were coming under scrutiny. A sub-committee proposed to the assembly of the university's graduates in Convocation in January 1878 that the university might be given special responsibility for education at the most advanced level, offering teaching in "those advanced branches of learning which can be best studied by men who have gone through a College course, and are no longer straitened by the successive prospects of examinations to be passed."[59] The subcommittee's report put the case for such a move strongly:[60]

> No years are more precious or more often wasted than those which intervene between graduation and active employment in professional life. It is at this time that the mind, disciplined by sound training and stored with acquired knowledge, is best able to enter upon special departments of pure science, and to advance knowledge by mature study and independent research. To aid in this object, to detain the graduate *inter silvas academi* and guide him in the pursuit of knowledge, seems peculiarly the office of a University as distinct from a College.

Despite the conviction with which the sub-committee commended the proposal, there could be no concealing the fact that implementation would create a demand for both teachers and facilities that would be difficult to realize, amid the financial constraints of the university. However, this did not rule out a compromise by which the university, while not necessarily offering instruction or supervision, would recognize a performance in re-

58. On this debate, see Percy Dunsheath and Margaret Miller, *Convocation in the University of London. The first hundred years* (London, 1958), 64–81.
59. "Report of the Annual Committee to Convocation at the meeting January 15, 1878," in *University of London Convocation. Standing Orders, Proceedings of Convocation (July 27th, 1877, January 15th and May 14th, 1878)*, 51. Membership of Convocation was open to all graduates of the University of London of three years' standing (the holders of certain degrees being admitted only two years after taking their degree).
60. Ibid.

search by internal or external students as a qualification for a higher degree.[61] Four months later, a motion to this effect, asserting "the importance of recognising independent research in the examinations for the higher degrees in such a way as the Senate may approve," duly went to Convocation and was approved, though with the proviso that before the resolution was implemented, the examiners who acted for the university should be consulted. Forty replies were received.[62] The great majority of these favored the institution of a D.Sc. by research, but doubts were expressed about what should be required. Even some of the most fervent advocates of the new degree insisted on the use of the term "independent research," rather than "originality," their point being that the degree should reward the acquisition of methods rather than the unrealistic pursuit of novelty.[63] This view prevailed. Almost without exception, the theses that resulted once the measure was put into effect contained no significant new departures: they reflected the regulation that a candidate might base his thesis "on the discovery of new facts observed by himself, or of new relations of facts observed by others."[64]

The provincial university colleges followed quickly in London's wake. The creation, in 1881, of up to four annual Bishop Berkeley Fellowships of £100 p.a. each, tenable for two years by students pursuing research, helped Owens College to get off to a fast start that benefited the sciences in particular: of the 43 fellows who received awards between 1881 and 1895 (when the scheme lapsed), only ten were in the humanities, compared with eleven in chemistry and five in physics.[65] By 1888, the hitherto informal position of research students at Owens had been regularized. Under new regula-

61. Ibid., 53.
62. Ibid., 63 and 73, in the minutes of the meeting of Convocation on 14 May 1878 and the report that the Annual Committee submitted to Convocation for the meeting. Among those who replied were Balfour Stewart, Carey Foster, Robert Clifton, Grylls Adams, John Tyndall (professor of natural philosophy at the Royal Institution and the Royal School of Mines), Isaac Todhunter (now a respected Honorary Fellow of St John's College, Cambridge), E.J. Routh (the most renowned of the private coaches who prepared candidates for the Cambridge Mathematical Tripos), John Burdon Sanderson (human physiology, University College, London), Edwin Ray Lankester (zoology, University College, London), G.D. Liveing (chemistry, Cambridge), and Henry Roscoe (chemistry, Manchester). The replies are in the archives of the University of London, Senate House, file RC 3/18 a–jj.
63. Interestingly, the only view unreservedly unfavorable to the idea of a research degree was expressed by one of the pioneers in the creation of teaching laboratories, Grylls Adams. His reply clearly indicates that he did not consider that the training of researchers could be carried out in the same way as teaching.
64. The phrase appeared annually in the *Calendar* of the University.
65. The fellowships were funded by W. Seaton Brown, a London friend of Thomas Ashton, the textile manufacturer and staunch champion of Owens College. How-

tions in that year, appropriately qualified students who were "desirous of undertaking original investigations" were allowed to use laboratory and library facilities for a payment of £3 per term, and in 1889 the Council went further by creating the position of Honorary Research Fellow.[66] Such a fellow, who had to be working in a scientific discipline, did not receive a stipend, but the distinction of the title and the indirect subsidy of being allowed to work free of charge in the college laboratories made the appointments attractive.

The gathering pace of change was eventually reflected in the institution of a scheme of national grants. In 1890 the Commissioners for the Exhibition of 1851 decided to spend some of their accumulated funds to create research scholarships to allow young graduates who had shown "distinct evidence of capacity for original research" to work in "some branch of Science, the extension of which is important to the national industries."[67] Despite the declared emphasis on industrially related knowledge, the conception of what constituted such knowledge was flexible enough to accommodate inquiries in "pure" science as well. The impact of the 1851 Scholarships was ensured by both their size and their numbers. An award of £150 p.a. or two years gave a generous period for study, and the decision to offer one every year to each of eleven universities and colleges throughout Great Britain (as well as, more intermittently, to certain other institutions) meant that the invigoration of intellectual life was felt across the system of higher education.[68]

---

ever, Brown's identity was not divulged until his death in 1897. On the fellowships and their recipients, see Philip Hartog, *The Owens College, Manchester (founded 1851). A brief history of the college and description of its various departments* (Manchester, 1900), 134–136, and Edward Fiddes, *Chapters in the history of Owens College and Manchester University, 1851–1914* (Manchester, 1937), 114–116 and 208–209.

66. *Owens College. Report of the Council to the Court of Governors* (Manchester, 5 Mar 1889), 17 (Appendix VI).

67. *Record of the Science Research Scholars of the Royal Commission for the Exhibition of 1851. 1891–1929* (London, 1930), 64–65.

68. Between 1891 and 1921 a grant was awarded annually to each of the following universities: Edinburgh, Glasgow, Birmingham (initially Mason College), Bristol (initially University College), Leeds (initially Yorkshire College), Liverpool (initially University College), London, Manchester (initially Owens College), and Sheffield (initially Firth College), and to Durham College of Science (later Armstrong College, Newcastle-upon-Tyne) and University College, Nottingham. A grant was awarded in alternate years to the Universities of St Andrews and Aberdeen, and one in turn to each of the three Welsh university colleges. In addition, a total of two grants were available for candidates from the Queen's Colleges of Belfast, Cork, and Galway and the Royal College of Science for Ireland in Dublin. The list is on page 66 of the *Record of the Science Research Scholars*, cited in the previous note.

Electricity and the physicists, 1880–1914

With a sensitivity bred of a quarter of a century of efforts to advance their discipline, experimental physicists were quicker than any other disciplinary group to see the advantages of the sudden extension of the study of electricity that occurred in the early 1880s. The extension, from the limited world of low currents in which they had been reared since the days of Wheatstone and Morse, propelled them into the new age heralded by the first international electrical exhibition in Paris in the autumn of 1881. Over a period of fourteen weeks the exhibition celebrated both the recent achievements and the potential of electrical power.[69] Almost 900,000 visitors to the Palais de l'Industrie in the Champs Elysées viewed displays of the latest electrical equipment, listened to the *théâtrophone* (which brought the music of the Opéra to telephone receivers in the Palais), and enjoyed an interior illuminated by the various competing systems of electric lighting. Outside the Palais, attractions ranged from the spectacle of the avenue de l'Opéra and other areas, lit for the first time by electricity, to the Siemens electric tram that plied a short route between the Palais and the place de la Concorde.

But the exhibition was, in both conception and realization, far more than just a popular success; its demonstration of the enhanced range of the applications of electricity stimulated intense commercial and technological interest and competition. While the battles raged on many fronts, that between the companies, entrepreneurs, and inventors engaged in promoting electricity for lighting was the fiercest and the one that attracted most public attention. A fundamental confrontation set the different types of arc-lamp, which had been developed and used with some success by Paul Jablochkoff, Jules Jamin, Dieudonné Lontin, Richard Werdermann, and others in the late 1870s, against the new incandescent lamp, which offered a gentler and more controllable or (as contemporaries put it) more "divisible" light appropriate for domestic and office use. Within each of these camps there were further confrontations, notably between the four exhibitors of incandescent light bulbs: two from the U.S. (Thomas Edison and Hiram Maxim) and two from Britain (Joseph Swan and St George Lane-Fox). Although the contest was close, it was Edison who emerged as the acknowledged leader and the dominant figure in electric lighting in Europe as he already was in the U.S. And it was he who, in the public's eye

69. For accounts of the exhibition, see François Caron and Catherine Berthet, "Electrical innovation: state initiative or private initiative? Observations on the 1881 Paris Exhibition," *History and technology, 1* (1984), 307–318, and Robert Fox, "Thomas Edison's Parisian campaign: incandescent lighting and the hidden face of technology transfer," *Annals of science, 53* (1996), 157–193.

and for many years to come, embodied the excitement and promise of the new age.

Edison's various enterprises, however, were not the only ones to prosper in the wake of the 1881 exhibition. At about the same time many other companies developed electrical interests on established foundations in mechanical engineering. Siemens & Halske in Berlin, Ganz in Budapest, and Mather & Platt in Manchester were just three such firms that now turned to the production of electrical machinery and equipment, mainly for lighting, and by the late 1880s, companies throughout the world were exploiting to very profitable effect a vogue for electricity that soon left virtually no sizeable town without some kind of central supply system. At this time, there was little correlation between the speed with which "networks of power" (to use Thomas Hughes's evocative term[70]) were constructed and the condition of the local economy. The cases of Spain and Finland show clearly enough that the implantation of a central generating station did not require the receiving country's economy to be in a highly industrialized, prosperous state; the vogue for electricity could make its mark anywhere.[71]

The struggles and mixed fortunes they had experienced since the 1850s had the effect of heightening physicists' perceptions of the benefits they might draw from these developments. Although, by 1880, electricity had come to be recognized as an important theoretical and experimental specialism within their discipline, the campaigns for laboratories and other facilities were seen as still having some way to go before the success which the most vociferous professors had claimed as their due was fully achieved. The conviction that much more remained to be done was voiced in communities of physicists everywhere, not least in the acknowledged pacemaker, Germany. There, as we observed earlier, a new emphasis on research as a qualifying criterion for academic appointments and a source of institutional prestige, apparent by the mid-century, had helped to promote the first moves towards the development of modern, independent physics institutes. Yet, despite some improvements, most notably at the University of Berlin (whose Physikalisches Institut was planned as early as 1871

70. The words form the main title of Hughes's influential book *Networks of power. Electrification in western society, 1880-1930* (Baltimore, MD and London, 1983), in which a systems approach is used in the analysis of the installations for the generation and distribution of electricity in Europe and the U.S. between 1880 and 1930.
71. Jordi Maluquer de Motes, "Los pionieros de la segunda revolución industrial en España: la Sociedad Española de Electricidad (1881–1894)," *Revista de historia industrial*, 2 (1992), 121–141, and Timo Myllyntaus, *Electrifying Finland. The transfer of a new technology into a late industrializing economy* (London and Helsinki, 1991), 13–58. Helsinki had a municipal generating station by 1884; see Myllyntaus, op. cit., 29 and 44.

and eventually opened in 1878), progress in the material provision for physics, in particular for original work, had been patchy, as it had been elsewhere in Europe.[72]

It is no coincidence that things began to change, and to do so decisively, from the 1880s. A main catalyst was the recognition, by physicists themselves and by their academic paymasters, of the challenges and opportunities arising from the escalating proliferation of the applications of electricity. Between 1882 and 1901, fourteen German universities built new physics institutes, many of them on an opulent scale and all of them incorporating substantial space and equipment for both teaching and research in electricity, as well as a supply of electricity for such auxiliary purposes as lighting, heating, and refrigeration.[73] As professors and university administrators agreed, the serious study of electricity had become an essential hallmark of a progressive, modern university. The cost of the new facilities was high—almost £200,000 for the fourteen institutes to which we have referred—but the perceived utility of physics meant that funds flowed freely, both from the state and other traditional sources and from an industry suddenly convinced of the commercial interest of the discipline. The point about utility as the main motor of disciplinary expansion was made unambiguously in 1896 by Friedrich Kohlrausch in an inaugural address to the Prussian Academy of Science in which he reflected on the causes of a remarkable expansion:[74]

> In the last few decades physics has grown out of its formerly outwardly very modest place into a recognized position of the first rank. It owes this, on the one hand, to its teaching, whose manifold power of cultivating thought and creativity no longer can be denied by anyone. However, if one were to analyze the motives which have led governments and legislatures to approve many millions for physical institutes, then the effective motive would prove to be the connection of physical research with life, with technology. . . .Physics has come to its rich resources through its interaction with cultural development, according to the modern principle, that one must risk capital in an enterprise that promises to be of use.

In other countries, the impact of the new applications of electricity had strikingly similar consequences for the growth of experimental physics after the preceding two or three decades of modest expansion. At University College and King's College in London, where short courses in telegraphy

72. Cahan, "The institutional revolution in German physics" (ref. 10), 12–15.
73. Ibid., 18–21.
74. Translated by Cahan, ibid., 57, from the *Sitzungsberichte der Königlich Preussischen Akademie der Wissenschaften, 33:2* (1896), 737.

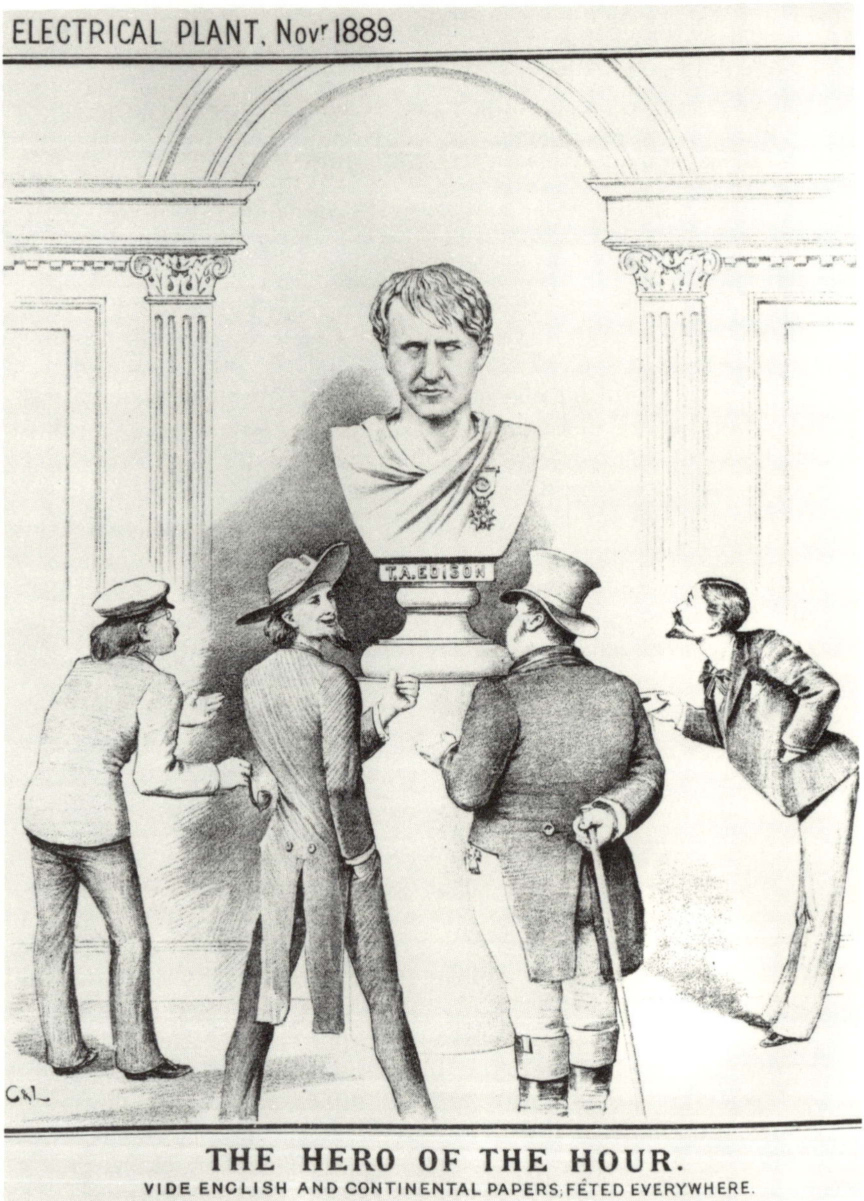

FIG 4. From the time Edison's name first became well known, in the late 1870s, he was idolized in Europe and North America as a symbol of the new age of electricity, not least because of the campaign of publicity, carefully orchestrated by the various companies bearing his name. From *Electrical plant*, Nov 1889. Reproduced with permission of the British Library (*shelfmark* Newspaper Library, 62).

and other aspects of low-current electricity had helped physicists to justify some expenditure on electrical apparatus as early as the 1860s, dramatic improvements occurred in the provision for electrical teaching and research after 1880. These allowed Carey Foster (at University College) and Grylls Adams (at King's) to turn their departments into leading centers for a practical form of instruction suitable both for young men going on into the electrical supply and manufacturing industries and for those who wished to proceed to original work at the interface between physics and electrical technology.

The alacrity with which Foster and Grylls Adams embraced high-current electricity as a proper domain for their departments illustrates the broad vein of pragmatism that ran through late-nineteenth-century physics. Another, unusually well documented example is that of Owens College. In a manner typical of the period, electricity had occupied only a minor part of the syllabus at Owens before 1880. But from the mid-1880s, under the influence of Arthur Schuster (the professor of applied mathematics and, from 1887, Langworthy professor of physics), the balance of activity in the physics department changed in ways that brought electricity unequivocally to the fore. In pursuit of his aim of heightening the importance of research and laboratory instruction, Schuster could build on the initiatives that his predecessor, Stewart, had taken since the early 1870s. But he could do so with an ambitiousness that would have been inconceivable in Stewart's day. The change was reflected equally in Schuster's conception of his new laboratory that was eventually opened in 1900 to replace the four dismal basement rooms that had been fitted out thirty years before for Stewart, and in a protracted restructuring of the syllabus that changed the focus of the discipline in Manchester.[75]

Schuster conveyed his position in repeated declarations that the traditional core of pure physics and the newer burgeoning world of practice should not be separated from each other. In his inaugural lecture as Langworthy professor, he presented his ideal of a physics laboratory and shop floor that were contiguous: it was, in his view, to the enduring benefit of both physics and industry that students going on to research and those, far more numerous, who aspired to industrial or teaching appointments should be taught in the same institution.[76] Nowhere was this unified preparation more appropriate and beneficial than in the case of electricity, where the association between science and engineering should be regarded as "little short of a necessity." It was in accordance with this principle that, as

75. On the laboratory and the work done in it, see *The physical laboratories of the University of Manchester. A record of 25 years' work* (Manchester, 1906).
76. Arthur Schuster, *Teaching in its relation to the industrial applications of science* (Manchester, 1889).

early as 1889 and in facilities that left much to be desired, Schuster converted Stewart's course in practical electricity (rooted in "special electrical and magnetic measurements," essentially the low-current work appropriate to telegraphy) into a far more ambitious one in "technical physics," subsequently and more accurately renamed "electrical technology." Here, under the guidance of a specialist lecturer (and former pupil at Owens College), William Haldane Gee, students were put through practical exercises with the clear purpose of inculcating immediately usable technical skills rather than the recondite intricacies of theoretical understanding.

The laboratory of 1900 was an exemplar of what had come to be regarded as a proper provision for physics by the turn of the century. It is true that the allocation of one of the three floors in the building entirely to electrical technology and electrochemistry bore the mark of Schuster's personal commitment, but professors of experimental physics in most other institutions would have shared his priorities in some degree. The main laboratory and a specially constructed annex, the John Hopkinson Memorial Wing, housed seventeen different examples of d.c. and a.c. generators and motors, an accumulator room (with two batteries with capacities corresponding to 300 and 135 ampère hours respectively), an electrical and magnetic testing laboratory, an alternating current laboratory, a standardizing room, and a drawing room—all for student use.[77] Although generous facilities also existed for spectroscopy, meteorology, heat, and work on high pressures, there could be no doubting where, in Schuster's conception, the heart of physics lay.[78]

Despite the benefit that physics laboratories everywhere drew from the impetus given by the rise of interest in electricity and its applications, local circumstances invariably acted to hasten or impede the process or to give it a peculiar twist. In France, for example, the new stimulus from electric-

77. *The physical laboratories of the University of Manchester* (ref. 75), 12–19.
78. There could be no doubting, of course, that Schuster remained first and foremost a physicist. His desire to preserve a close connection between the scientific and technological aspects of electricity subsided when it became clear that the Manchester Technical School was expanding its provision for the teaching of electrical engineering. In fact, it was Schuster himself who, on his retirement, spoke up as strongly as anyone for the appointment of Ernest Rutherford to his chair, knowing that this would open a new phase in the development of the physics department at Owens College, reducing its involvement in electrical engineering and bringing it closer to the advanced frontiers of experimental physics. See Robert H. Kargon, *Science in Victorian Manchester. Enterprise and expertise* (Manchester, 1977), chapter 6, and Anna Guagnini, "The fashioning of higher technical education and training in Britain: The case of Manchester, 1851-1914," in Howard F. Gospel, ed., *Industrial training and technological innovation. A comparative historical study* (London, 1991), 69-92.

FIG 5. The new physics laboratory of Owens College, Manchester, was formally opened by Lord Rayleigh in 1900. The cost of the building and equipment, including the John Hopkinson Memorial Wing (named after John Hopkinson, who had been a student at Owens, and devoted entirely to electrical engineering) stood at over £36,500 by 1906. For a break-down of the source of the funds, which came almost entirely from private donations, see *The physical laboratories of the University of Manchester. A record of 25 years' work* (Manchester, 1906), 127-129. The plate reproduced here appears as a frontispiece in this volume. Reproduced with permission of the London Library.

ity in the 1880s coincided with a period when the national University, of which the faculties of science, like all other faculties, were a part, enjoyed an unprecedented degree of official favor, mainly for political reasons. In the eyes of the leaders of the Third Republic, consolidated in power after a series of elections that definitively advanced the interests of the Left in the later 1870s, the University symbolized a progressive structure for teaching and research that squared well with outgoing modernizing ideals, in contrast with the traditionally more conservative world of the Ecole Polytechnique and the other technical *grandes écoles*, which most republicans regarded with distinctly less favor.[79] The life of the faculties was also

79. Weisz, *The emergence of modern universities in France* (ref. 22), esp. chapter 3, and, for briefer accounts, Weisz, "Le corps professoral de l'enseignement supérieur" (ref. 22), 202–232, and "The French universities and education for the new professions, 1885–1914: an episode in French university reform," *Minerva, 17* (1979), 98–

FIG 6. The dynamo house in the new physics laboratory at Owens College, Manchester, was part of the John Hopkinson Memorial Wing. Some of the machinery, specially designed and constructed for educational purposes, was presented by the Manchester engineering firm Mather & Platt. From *The physical laboratories of the University of Manchester. A record of 25 years' work* (Manchester, 1906), facing p. 14. Reproduced with permission of the London Library.

crucially affected by the fact that ministerial policies in the 1880s inaugurated a wave of academic devolution. In this, the faculties were allowed, indeed urged, to seek industrial and other local support to augment the core provision (mainly of salaries) that continued to come from the Ministry of Public Instruction.

The resulting union of national and regional interests helped to encourage a spate of university building throughout France. Although faculties of all types derived some measure of benefit, those of science and medicine enjoyed particular favor, drawing substantial funds from what was usually a mixed constituency of public and private benefactors. Between 1885 and 1900, in the provinces alone, industry, individual donors, and departmental authorities contributed well over £1 million (roughly three-quarters of the total) for the construction and fitting out of faculties of science.[80] But in science as in all disciplines, the most lavish provision was made in Paris. There, Nénot's plan for the "new Sorbonne," elaborated in the mid-1880s and realized, with relatively few changes, between then and 1900, allocated 9200 sq. meters of the total floor space of 21,000 sq. metres to the Faculty of Science.[81] Although in the provinces the scale was smaller, the favor shown towards science (with physics and chemistry always very much to the fore) was just as marked.[82] The most spectacular showpieces were those in Lyon, Bordeaux, Nancy, and Lille, but no science faculty was untouched by the movement; indeed, the provision made for one faculty, in Toulouse, was so lavish as to provoke unfavorable local comment on the grounds that it was out of proportion to the number of students.[83] Extravagant or not, however, the new facilities served as a powerful stimulus, and in every faculty of science the mutually reinforcing interactions between material investment and activity in research made their mark. As Terry Shinn has shown, in the last quarter of the century, the number of both academic posts and publications produced *per capita* in science increased

128. On republican attitudes towards the Ecole Polytechnique, see Terry Shinn, *Savoir scientifique & pouvoir social. L'Ecole polytechnique, 1794–1914* (Paris, 1980), chapter 4.
80. Terry Shinn, "The French science faculty system, 1808–1914: institutional change and research potential in mathematics and the physical sciences," *Historical studies in the physical sciences, 10* (1979), 271–332, on 311–312.
81. Nénot, *Monographie de la nouvelle Sorbonne* (ref. 34), 20.
82. At Bordeaux, for example, the floor area available for science in the new building of the faculties of science and letters (inaugurated in 1885) was approximately a third of that in the "new" Sorbonne. See G. Rayet, "Histoire de la Faculté des Sciences de Bordeaux (1838–1894)," *Actes de l'Académie Nationale des Sciences, Belles-Lettres et Arts de Bordeaux,* 3e série, 59e année (1897), 5–388, on 50–54.
83. Paul Sabatier, "La Faculté des Sciences," in *1229–1929. L'Université de Toulouse. Son passé–son présent* (Toulouse, 1929), 189–209, on 193–194.

FIG 7. The research laboratory for physics in the "new" Sorbonne, directed at the time of this illustration by Gabriel Lippmann. From *La Nature*, 26e année, 1er semestre (1898), 225. Private collection.

FIG 8. The machine room of the research laboratory for physics in the "new" Sorbonne, showing one of the room's three dynamos. From *La Nature*, 26e année, 1er semestre (1898), 225. Private collection.

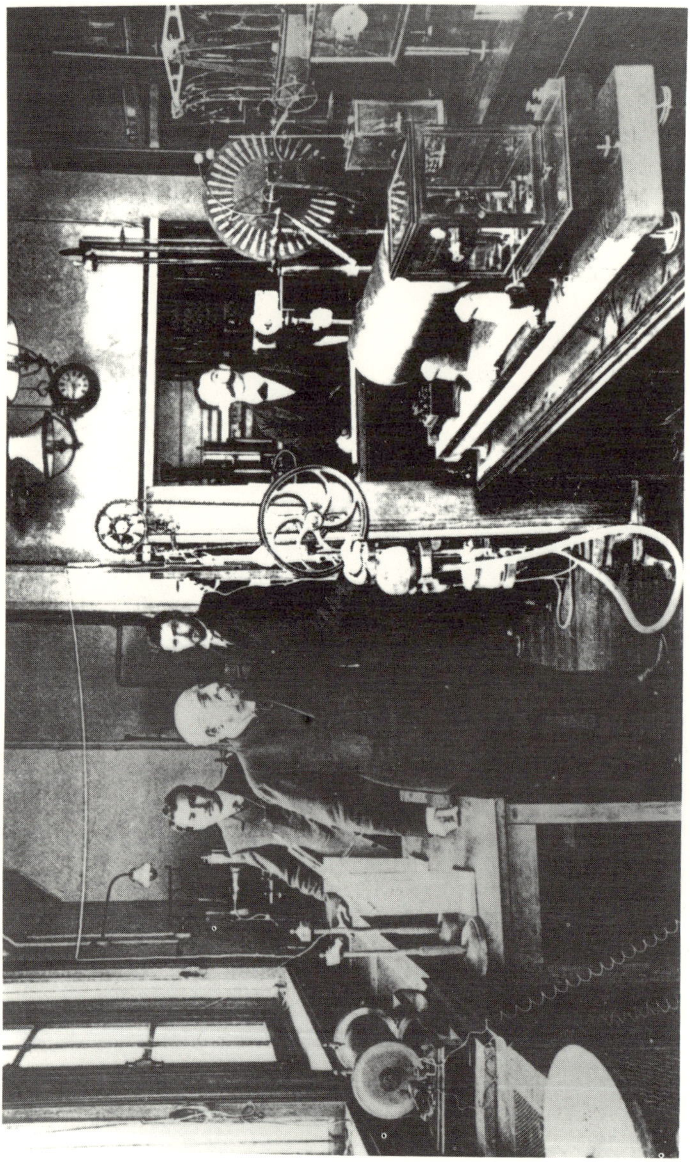

FIG 9. Henri Pellat with electrical apparatus in the physics laboratory of the "new" Sorbonne, c. 1900. Pellat, who worked mainly on electricity and gas discharge phenomena, held a specially created chair of physics at the Sorbonne from 1899 until his death in 1909. Reproduced with permission of the Service du livre ancien, Bibliothèque de la Sorbonne, Paris, from a postcard (RBA 163 in-f°, f.67, n°2).

to unprecedented levels, and research (beyond what was required for the doctorate) became a leading criterion for appointments.[84]

The case of France illustrates the benefits that favorable political and ideological circumstances—a combination of an intellectually exciting and potentially useful discipline developing in a period of growth in the university system—could yield for a community of physicists that was ready to exploit them. Where such a combination of circumstances did not exist, similar advances were far less easily achieved, even impossible. In Italy, in fact, the turn of the century was marked by, if anything, a step backwards in the already shaky position of the main experimental sciences, both physics and chemistry. Physicists in Italy were as aware as their peers anywhere of the progress that was being made in the study of electrical phenomena, and some of the most distinguished of them, including Augusto Righi, Pietro Blaserna, and Orso Maria Corbino, turned eagerly to what they recognized as a leading-edge specialty.[85] However, their work in the field never broke completely with the elaboration of what were essentially known phenomena. This is not to say that their contributions were insignificant. In accurate measurement and other work involving detailed observation, they showed real competence, but they constantly followed in the wake of their more brilliant colleagues in other countries, to the point of being overawed by what was achieved abroad.

Contemporaries and historians alike have highlighted the link between the stagnation of the experimental sciences in Italy and the lack of adequate funding—a lack that is properly ascribed to the secondary position that these sciences occupied in Italian higher education and continued to occupy well into the new century. As far as physics is concerned, one of the contributory underlying weaknesses lay in the failure of most Italian physicists to exploit the "applied" nature of their discipline, in particular of electricity, as a bargaining counter. The majority of the experimental physicists who might have used this argument remained determinedly committed to non-utilitarian values and, with few exceptions (of whom Antonio

84. Shinn, "The French science faculty system" (ref. 80), 302–315, and Paul, *From knowledge to power* (ref. 30), 251–287.

85. This sense is conveyed in Orso Maria Corbino, "Il contributo italiano ai progressi della elettrologia nell'ultimo cinquantennio," in *Atti della Societa Italiana per il Progresso delle Scienze, V riunione. Roma 1911* (Rome, 1912), 275–306. The main features and limitations of Italian physics in the period are examined in Barbara J. Reeves, "Le tradizioni di ricerca della fisica italiana nel tardo diciannovesimo secolo," in Vittorio Ancarani, ed., *La scienza accademica nell'Italia post-unitaria. Discipline scientifiche e ricerca universitaria* (Milan, 1989), 1–36.

Pacinotti was the most notable),[86] took a visible pride in the distance that separated them from direct involvement in the commercial exploitation of their science. As we observe in Chapter 3, it was left for teachers in the university-level engineering schools, the *scuole di applicazione per ingegneri,* to reap the benefits of the mounting interest in the industrial applications of electricity, although their efforts did little to advance the interests of research.[87]

It is not hard to identify the circumstances that contributed in Italy to the physicists' lukewarm response to the rhetorical opportunities offered by the boom in electricity and its applications. Governmental indifference to science and to the laboratories and other facilities (including the opening of academic positions) on which its welfare depended certainly weighed heavily.[88] But equally discouraging were the structure and traditions of Italian industry: even in those fields, such as electrical engineering, in which

86. Soon after graduating in physics at the University of Pisa in the mid-1860s, Pacinotti (who went on to become professor of physics at the University of Palermo) devised a machine, with a distinctive iron-ring armature, that could function both as a generator and as a motor and was later incorporated as a core component of the dynamo. Although Pacinotti himself did not succeed in the commercial exploitation of his invention, he maintained a keen interest in the industrial applications of electricity throughout his life. See Giovanni Polvani, *Antonio Pacinotti. La vita. L'opera* (Milan, 1932).
87. Despite the broad distinction we make here, it is important not to lose sight of the diffuseness of the boundary between the communities of experimental physicists and electrical engineers in Italy in the late nineteenth and early twentieth centuries. Augusto Righi, for example, graduated in engineering before going on to a chair of physics at the University of Bologna, and Galileo Ferraris, the professor of electrical technology at the Scuola di Applicazione per Ingegneri, later to become the Politecnico, of Turin, maintained a strongly theoretical approach even in his most applied research.
88. In the last years of the century, the opening of the new laboratory of the Istituto Fisico of the University of Rome "La Sapienza" in 1899 was the only significant move towards the provision of a modern, well-equipped environment for experimental research. The government's provision of over £25,000 for the building and its equipment was strikingly, if not spectacularly, generous. But despite the quality of the facilities, the laboratory took off haltingly and was slow to generate the flow of new research that its founder and first director Pietro Blaserna hoped it would. For evidence of the pride that accompanied its opening, see the calendar of the University of Rome, *Annuario 1899–1900* (Rome, 1899), Appendix, 14–19. And for a modern comment, see Reeves, "Le tradizioni di ricerca della fisica italiana" (ref. 85), 62–63. The relations between funding and research in Italian physics are well discussed in Silvana Galdabini and Giuseppe Giuliani, "Physics in Italy between 1900 and 1940: the universities, physicists, funds, and research," *Historical studies in the physical and biological sciences, 19* (1988), 115–136.

entrepreneurial activity was brisk, manufacturing depended heavily and quite profitably on foreign imports and technology and hence did little to encourage the closer participation of the scientific community.[89] Deeply ingrained institutional constraints within the university system also played an inhibiting role.[90] The main constraint was the perception of university-level education as first and foremost a means of preparing young men for professional careers. The cultivation of experimental research, by contrast, was an individual pursuit of the professoriate and it remained a marginal activity, lacking both adequate material support and formal recognition as part of university life, for much longer than it did in other countries. Governmental grants for research were rare, and the few students who obtained them tended to use them for work in foreign universities, a growing number of which, as we have seen, offered not only training in experimental practice but also the reward of a doctorate, a degree that did not exist in the Italian system until well after the second world war. A comparison between the laboratory-based training that was available in the leading European and North American universities by the turn of the century with the few and rather precarious lecture courses, the *corsi di perfezionamento*, which were all that Italian universities provided at the postgraduate level, confirms the wisdom of the decision to work abroad, whenever that opportunity was available.

Finally, a word on the balance between supply and demand. The sense of excitement that surrounded experimental courses and the institutional and intellectual momentum that went with it, would have been hard to sustain if there had been any shortage of enrollments. But the ubiquity of the developments we have described as accelerating from about 1880 was matched by a corresponding shift in the preferences of students. The case of Owens College illustrates the point. There, the number of students enrolled in the physics laboratory rose from eight in 1871/2 to 62 in 1885/6, 105 in 1890/1, and finally to 183 in 1895/6. The expansion gave the college undisguised satisfaction despite the fact that by no means all of the students wished to specialize in physics, still less in Schuster's favored specialism of electricity.[91] A similar expansion occurred in Germany, sustained largely by recruitment from students preparing for careers as

89. See our own comments on this below, in Chapter 4, and the detailed study by Roberto Maiocchi, "Il ruolo delle scienze nello sviluppo industriale italiano," in Gianni Micheli, ed., *Storia d'Italia. Annali*. Vol. 3. *Scienze e tecnica nella cultura e nella società dal Rinascimento ad oggi* (Turin, 1980), 863–999.

90. Vittorio Ancarani, "Università e ricerca nell'Italia post-unitaria. Saggio introduttivo," in Ancarani, *La scienza accademica nell'Italia post-unitaria* (ref. 85), 1–36, esp. 29–35.

91. *The physical laboratories of the University of Manchester* (ref. 75), 1.

doctors, pharmacists, industrial chemists, or in school teaching, rather than as physicists.

Despite this diversity of career aspirations, as attendances at lectures and laboratory classes in physics increased, so necessarily did the exposure of students to the electrical content of the courses. One consequence was a small but not insignificant growth in the number of pupils who wished to move on from normal undergraduate coursework to research in electricity. The process was helped by the improved provision that the more ambitious universities had made, by now, for research students. The laboratories of the University of Cambridge (in biology and physiology as well as in physics) emerged as particularly attractive,[92] not least because of the decision, taken in 1895, to recognize the degrees held by graduates of other universities who wished to enter the university to undertake research. But across Europe and North America (especially at the Johns Hopkins University in Baltimore, where physics had developed strongly since the late 1870s under Henry Rowland), the signs of a will to establish research traditions in particular areas of science and thereby to win students from other universities and even other countries were by no means rare.[93]

Physics prospered conspicuously through this movement. At the postgraduate, as at other levels, it drew on perceptions of it as at once intellectually challenging and potentially useful, a discipline that, especially in electricity, stood at the cutting edge of modernity. Yet by the turn of the century, just as physicists were beginning to enjoy the opportunities afforded by the disciplinary expansion and public attention for which they had struggled for decades, they could not avoid the uncomfortable recognition of unanticipated but crucial limits to their new prosperity. These limits arose from competing disciplinary aspirations that were beginning to make electricity an attractive field for exploitation by engineers as well. From the time the first departments or special sections of electrical engineering were created in the 1890s, the status of physics departments as the main academic settings for work in applied electricity became a potential source of friction similar to that which occurred at about the same time on the boundaries between chemistry and chemical engineering. And potential for conflict soon became a reality. What the electrical engineers re-

92. The rise of Cambridge as a force in laboratory-based research in biology and physiology is a central theme in Gerald L. Geison, *Michael Forster and the Cambridge school of physiology. The scientific enterprise in late Victorian society* (Princeton, NJ, 1978).
93. The significance of the growth of laboratories and of the recruitment of advanced students to work in them is brought out in several contributions to Gerald L. Geison and Frederic L. Holmes, eds., *Research schools. Historical reappraisals*, published as *Osiris*, 2nd ser., *8* (1993). See, for example, Jack Morrell, "W. H. Perkin, Jr., at Manchester and Oxford: from Irwell to Isis," ibid., 104–126.

sented, above all, was the physicists' now standard rhetoric about the unique utility of their particular skills and approach. As the engineers correctly perceived, such rhetoric set them and physicists in competition with one another and could do nothing but harm to their own campaign for the creation of laboratories and the advancement of their discipline. It called for and in the two decades before the first world war duly provoked a new assertiveness that served to bind the nascent community of electrical engineers and give it an unprecedented sense of its worth and potential.

3 ELECTRICAL ENGINEERING

As the decade in which electricity moved firmly and quite suddenly to the forefront of physics, the 1880s marked an important turning point in the history of the discipline. Physicists responded with enthusiasm to the challenge and the new opportunities, and continuing technological advances ensured that the sense of excitement was maintained throughout the decade and even enhanced. But by the early 1890s the changes, now focussed mainly on the use of alternating current at voltages far greater than any that had been produced by even the most powerful direct current generators, were tending to give a greater stimulus and scope to the engineering disciplines than they did to physics. It was the new breed of educated electrical engineers who tackled the complex problems faced in the scaling up and expansion of the use of electricity and who produced most of the improvements that transformed the technology between 1890 and 1914. Not surprisingly, this was the period in which electrical engineering became established as the most exciting and intellectually sophisticated specialty in higher technical education. Important elements in this were the "scientific" image of the discipline, bred of the recent developments in electromagnetism, and the integration, from the start, of laboratories in its pedagogical practices. Yet, even with these advantages, it was by no means easy to establish the ideals and practices of research as normal activities of higher technical schools.

The magnitude of the technological shift resulting from the exploitation of alternating current and higher voltages became abundantly clear at the international electrical exhibition of 1891 in Frankfurt-am-Main. By now, the performance of the early a.c. generators and transformers constructed for lighting plants by Ganz, Siemens, and (in the U.S.) Westinghouse had been far surpassed in a series of improvements, affecting both output and reliability, which at last had made these sources of electricity a viable alternative to the established d.c. dynamos. The core achievement lay in innovations in design that had led to the construction of ever larger generators and a display of machinery at Frankfurt that eclipsed its predecessors. Even the daringly original arrangement of generators, cables, and trans-formers that from 1889 had allowed Sebastian Ziani de Ferranti to attempt the supply of alternating current at 10,000 volts from the power station at Deptford, on the Thames to the east of London, was surpassed.[1]

The most captivating demonstration of the new technology at Frankfurt was in the field of long-distance power transmission—an essential area for development if electricity was to be used in mass transport and the provision of power for manufacturing. Nine years earlier, at the international electrical exhibition in Munich, the French engineer Marcel Deprez had mounted one of the first public demonstrations of transmission between Miesbach and the main exhibition hall in Munich, 57 km away, where the power was used to drive a centrifugal pump feeding a small waterfall (see fig. 10). Deprez, a mechanical engineer who had developed an interest in electricity in his thirties, had used direct current and a small dynamo capable of an output of 1.5 kW at no more than 2000 volts (see fig. 11).[2] But in 1891 Oskar von Miller, who was responsible for a demonstration similar to Deprez's (even to the point of using the transmitted electricity to power a waterfall), was able to transmit alternating current not only over a much longer distance (from Lauffen, 175 km from Frankfurt) but also at a voltage of 25,000 volts that allowed losses along the line to be greatly reduced.[3]

It is beyond question that the success of the Frankfurt demonstration of 1891 owed a great deal to von Miller's personal vision and entrepreneurial ability.[4] But the achievement also rested on a broader technological base, contributed in the main by two prominent electrical firms that took a lead-

1. Ferranti began designing the plant in 1886, in his capacity as Chief Engineer of the London Electric Supply Corporation. His aim was to supply lighting plants in the center of London from a power station of unprecedented size situated at Deptford on the eastern outskirts of the city. By 1889, high-voltage alternating current was being transmitted by means of 28 miles of underground cables. However, technical problems as well as unfavorable political and commercial circumstances eventually compelled him to abandon his project. For an analysis of Ferranti's pioneering plan and the causes of its failure, see Percy Dunsheath, *A history of electrical engineering* (London, 1962), 159–168; Thomas P. Hughes, "The bold conception of S.Z. de Ferranti," *Consulting engineering, 3* (1963), 75–79; Arthur Ridding, *S.Z.de Ferranti. A brief account of some aspects of his work* (London, 1964), 12–19; and Thomas P. Hughes, *Networks of power. The electrification of western society 1880–1930* (Baltimore, Md. and London, 1983), 227–261.

2. Wilhelm von Beetz et al., eds., *Offizieller Bericht über die im Königlichen Glaspalaste zu München 1882...stattgehabte Internationale Elektricitäts-Ausstellung* (Munich, 1883), 101–106.

3. *Offizieller Bericht über die Internationale Elektrotechnische Ausstellung in Frankfurt am Main 1891* (2 vols., Frankfurt, 1894), *1*, 375–388, and *2*, 319–451. See also, with special reference to the contribution of the Oerlikon company, *Maschinenfabrik Oerlikon (Zürich). Electro-technische Ausstellung in Frankfurt a/Main* (Zurich, 1891).

4. Von Miller studied civil engineering at the future Technische Hochschule of Munich in the early 1870s and began his career as an apprentice civil engineer in the Bavarian state service. The Munich exhibition in 1882, where he witnessed

ing role in the design and construction of the experimental plant. These firms were the Swiss Maschinenfabrik Oerlikon, with its director P.E. Huber-Werdmuller (a former student of the Eidgenössische Technische Hochschule in Zurich) and gifted technical director Charles E.L. Brown, and the Berlin-based AEG, headed by Emil Rathenau (also a graduate of the ETH) and with an outstanding chief engineer, Michael Dolivo Dobrowolsky, a graduate of the Darmstadt Technische Hochschule who had begun experimenting on polyphase motors soon after Ferraris and Tesla. The alliance, which brought together complementary experience and know-how in what was still an uncharted and risky area of technology, was a powerful one and it provided von Miller with ample scope in his aspiration to lead in the techniques of transmission.

The changing face of engineering: From workshop to laboratory

The crowds at the Frankfurt exhibition who admired von Miller's waterfall and the 1000 lamps that were lit with the power from Lauffen perceived that a new age had dawned. The point was even clearer for physicists, who recognized that their contribution to applied electricity would have to evolve with the technology on show at Frankfurt if they were to maintain the strong position they had established during the 1880s. Through that decade and on into the early 1890s, they repeatedly demonstrated their capacity to adapt and thereby the continued value of a degree in science, preferably with a strong component of electricity, as a disciplinary foundation for someone wishing to work in the new field. The cases of the brothers John and Edward Hopkinson, who studied physics and mathematics at Owens College and then took high positions as wranglers in the Mathematical Tripos at Cambridge before embarking on distinguished careers as electrical engineers, illustrate this degree of continuity well. The sons of an engineer who was a partner in a Manchester-based company, both John and Edward were determined to follow in their father's footsteps, though only after they had undergone some relevant formal instruction.[5]

Deprez's demonstration, inspired his interest in electrical technology. Seeing how the demonstration might be developed for the exploitation of the abundant water power of the Alps, von Miller devoted a decade to the realization of his aim, both as the head of his own consulting firm and as manager of Emil Rathenau's newly established Deutsche Edison Gesellschaft (soon to become the AEG) from 1883. In a large literature on von Miller's life and work, see esp. his autobiographical sketch, covering his career until the year of the Frankfurt exhibition, in Walther von Miller, ed., *Oskar von Miller nach eigenen Aufzeichnungen, Reden und Briefen* (Munich, 1932), 10–65, esp. 55–65 on the transmission between Lauffen and Frankfurt.

5. The work of the Hopkinson brothers has yet to receive the attention it deserves. The only sources are a memoir by John Hopkinson's son, Bertram, published in the

FIG 10. The generating station at Miesbach, where a steam engine was used to produce direct current which was transmitted over a distance of 57 km to Munich. The waterfall that the current powered (see fig. 11) was one of the main popular attractions, as well as one of the technical successes, of the international electrical exhibition of 1882. From Frank Geraldy, "Le transport électrique de la force," *La lumière électrique, 10* (1883), 361-368, on 366. Private collection.

FIG 11. The waterfall in the Glass Palace, Munich, using water raised by power transmitted from Miesbach (see fig. 10). From Frank Geraldy, "Le transport électrique de la force," *La lumière électrique, 10* (1883), 361-368, on 367. Private collection.

FIG 12. Polyphase dynamo, constructed by Oerlikon, at Lauffen, 1891. Alternating current from this dynamo was transmitted at a voltage of 25,000 volts to the hall of the international electrical exhibition in Frankfurt-am-Main, 175 km away. From *Offizieller Bericht über die Internationale Elektrotechnische Ausstellung in Frankfurt-am-Main 1891* (2 vols., Frankfurt, 1894), *1*, 381. Reproduced with permission of the Institution of Electrical Engineers Archives.

In the event, the education of John and Edward Hopkinson took a longer and intellectually more elevated route than they had originally intended. John attended Owens College from 1865 to 1867 and then went to Cambridge to read for the Mathematical Tripos, emerging as Senior Wrangler in 1871; in the same year he took the degree of D.Sc. of the University of London (by examination) in applied mathematics and in Cambridge was elected a Fellow of Trinity College. His younger brother Edward also studied at Owens College, from 1874 to 1877, moved on to Cambridge, where he was tenth Wrangler in 1881, and in the same year took the examination for the degree of D.Sc. of the University of London (in the branch of electricity treated mathematically). Clearly, the Hopkinsons would have assimilated little at Owens College and Cambridge that bore directly on the kind of problems they were to face in their careers.

---

first volume of John Hopkinson, *Original papers* (2 vols., Cambridge, 1901), ix–lxvi, and James Greig's brief *John Hopkinson. Electrical engineer* (London, 1970). There is even less on Edward Hopkinson.

FIG 13. Motor at the international electrical exhibition in Frankfurt-am-Main, 1891. Powered by electricity from Lauffen (see fig. 12), the motor raised water for the creation of an artificial waterfall similar to the (smaller) one, displayed at the Munich electrical exhibition of 1882, that is shown in fig. 11. From *Offizieller Bericht über die Internationale Elektrotechnische Ausstellung in Frankfurt am Main 1891* (2 vols., Frankfurt, 1894), *1*, 383. Reproduced with permission of the Institution of Electrical Engineers Archives.

But they took steps to marry their scientific background with extensive experience gained on the job. John, after a brief apprenticeship at his father's firm, joined the optical glass factory of Chance Brothers & Co. in Birmingham, as engineer and manager of the lighthouse and optical glass departments, and then moved permanently into the field of electrical engineering under the personal influence of William Thomson and through his reading of Maxwell's *Treatise on electricity and magnetism*. In 1877 he left the company and set up as a consulting engineer in London; there he quickly acquired a high reputation and continued to practice as a consulting engineer even after his appointment as professor of electrical engineering at King's College, London, in 1890. Edward, for his part, was apprenticed to William Siemens, and in 1884 entered the Manchester-based mechanical engineering firm of Mather and Platt, then just moving into the booming field of electricity (see fig. 14). There he took charge of the electrical engineering department and worked with John on the design of an improved form of Edison's dynamo, whose manufacturing rights were held by Mather and Platt.

The careers of the Hopkinsons show how even at the end of the century graduates in physics could develop an engineering approach in their professional engagement with the design, production, and testing of electrical machinery. By now, however, and even more markedly by 1914, this kind of professional retooling was becoming less common. Everywhere, electrical engineering was firmly established as a discipline in its own right, with a distinct *cursus* of study and examinations and a style that bore the marks of its origins as an offshoot not of physics but of the highly practical world of engineering. The background to this lies in the transition through which industrially oriented technical education achieved university-level status. An essential facet of the transition was the growing importance that came to be attached to laboratory-based instruction in the syllabuses of higher technical schools. In certain respects, the schools were complying with the trend towards a greater role for laboratory exercises and research in the curricula of higher education. However, the motivations and circumstances that underlay the process in the technical schools were very different from those that led to the opening of science laboratories, in particular of physics laboratories, in the universities.

By the 1860s, higher technical education had already started on the long road towards full academic recognition, chiefly through the fashioning of new styles of teaching and professorial activity.[6] One virtually universal

6. An excellent comparative study of the rise of "school culture" in engineering in Europe and the U.S. is Peter Lundgreen, "Engineering education in Europe and the U.S.A., 1750–1930: The rise to dominance of school culture and the engineering professions," *Annals of science*, 47 (1990), 33–75. Lundgreen explores very helpfully

change, in both Europe and the U.S., was in the balance between workshop-based training and training in laboratory techniques, with the latter slowly but steadily assuming more importance. It cannot be stressed too strongly, however, that, despite the shift, workshop training was a tenacious survivor. Where a commitment to practical instruction had particularly strong roots, workshop skills maintained a respected place in the syllabus. This was true, for example, of King's College in London (whose workshops were described by the Royal Commission on Technical Instruction in 1884 as the most remarkable feature of the institution[7]), the French middle-level technical schools, the *écoles d'arts et métiers* (whose *ateliers* were regularly upgraded),[8] and the Polytechnisches Institut (later the Technische Hochschule) in Vienna, which made a principle of its support for workshop teaching.[9]

By the 1880s, however, the value of the school-based workshop tradition was under scrutiny, and in places under fire. As might be expected, it was criticized by the more academically minded members of the professoriate. But it also had critics among those of a more practical disposition, who regarded the workshop facilities in most schools and departments of engineering as a poor, even misleading microcosm of a true industrial environment. Even in Germany, where the move towards a greater emphasis on science in the syllabuses of the Technische Hochschulen was beginning by 1870 to arouse a significant backlash from engineers close to the world of industry, the old-fashioned school workshop was no longer regarded as the best antidote to excessive abstraction.[10]

The criticism of school-based workshops was the culmination of a dual development that had gathered strength since the 1850s. One strand was the progress in the theoretical foundations of engineering, to which we referred in chapter 1. The other, closely related and bearing more directly on the theme of this chapter, was the growth of investigations of an experi-

---

the role of state bureaucracies (as opposed to the needs of industry) as a leading cause of the different pace at which formal engineering education was developed in different countries in the nineteenth century.

7. *Report of the Royal Commission on Technical Instruction* (London, 1882–1884), second report, *1*, 396.

8. Charles R. Day, *Education for the industrial world. The Ecoles d'Arts et Métiers and the rise of French industrial engineering* (Cambridge, MA, and London, 1987), 99.

9. Alfred Lechner, *Geschichte der Technischen Hochschule in Wien (1815–1940)* (Vienna, 1942), 75–195, and Hedwig Gollob, *Geschichte der Technischen Hochschule in Wien nach neugefundenem Aktienmaterial bearbeitet* (Vienna, 1964), 49–79.

10. Kees Gispen, *New profession, old order. Engineers and German society, 1815–1914* (Cambridge, 1989), esp. chapter 6.

FIG 14. A montage of products by Mather & Platt, prepared for the Manchester Royal Jubilee Exhibition of 1887. This illustration reflects the growing prominence of electrical machinery in the company's output. From Loris Emerson Mather, *The Right Honourable Sir William Mather, P.C.,. LL.D., M.Inst.C.E., 1838-1920* (London, 1925), facing p. 16. Reproduced with permission of the London Library.

mental character that had become a feature of engineering during the third quarter of the century. Generally, in this period, such investigations were pursued by individuals who, even if they held academic posts, performed their work outside the institutional sphere of universities and higher technical schools. Most of the experiments were carried out on behalf of large private companies, such as railway companies, or of public agencies, notably the army or navy—organizations that were particularly responsive to economic and social pressures directed at efficiency, cost reduction, and safety standards.

An early example was the work of Eaton Hodgkinson, a self-taught engineer whose achievements in professional practice and experimental research led to his election as a Fellow of the Royal Society and his appointment, in 1847, as professor of mechanical engineering at University College, London. The seminal investigation of the strength of materials that Hodgkinson conducted from 1822 until his death in 1861 was largely based on experiments carried out in the very practical environment of William Fairbairn's works in Manchester and Millwall; likewise, it was the design of the Britannia and Conway tubular bridges in the late 1840s that provided the stimulus for an elaborate series of tests on the mechanical properties of iron beams and tubes conducted with Robert Stephenson and Fairbairn.[11] In a similar way, between 1858 and 1861, David Kirkaldy, a London consulting engineer, worked on the mechanical properties of wrought iron and steel beams for Robert Napier and Son; the main aim of the experiments, which Kirkaldy performed on specially designed testing machinery in his private laboratory in Southwark, was to provide the Napier company with information relevant to the construction of high-pressure boilers and marine engines.[12] In continental Europe, too, such practices became more common. In the 1850s, for example, Adolphe Hirn, a textile manufacturer in the Mulhouse region of southern Alsace, launched an important program of tests on working steam engines in his factory. It was one of these tests, conceived with a mixture of scientific and practical motives, that confirmed the principle of the conservation of energy by demonstrating for the first time that the heat expelled from an engine was less

11. For a brief biography of Hodgkinson, see *Proceedings of the Royal Society, 12* (1862–1863), xi–xiii. The results of the experiments arising from Hodgkinson's involvement in the work on the Conway and Britannia bridges were published in Edwin Clark ("with the sanction, and under the supervision, of Robert Stephenson"), *The Britannia and Conway tubular bridges* (2 vols. plus Atlas, London, 1850). The experiments were financed jointly by the Royal Society and Stephenson himself.
12. David Kirkaldy, *The result of an experimental inquiry into the tensile strength and other properties of various kinds of wrought iron and steel* (Glasgow, 1862).

than that entering it by an amount proportional to the amount of work obtained.[13]

The work of these individuals demonstrated that the introduction of sophisticated testing instruments and rigorous measuring procedures was capable of benefiting industry. In the process, it also fostered the image of mechanical engineering as a discipline in which experimental methods as well as theory had their place. As a result, through the 1860s and 1870s a new generation of engineers began to come to maturity with a preparation that embodied these developments. It was a preparation in which theory was not derived simply from science but equally from an emerging corpus of experimental evidence gained primarily on the site. Crucial to this development was a small number of leaders in academic engineering who exerted influence both through their own teaching and by the textbooks they wrote, some of which enjoyed immense success over long periods. In continental Europe, Ferdinand Redtenbacher was in this respect a pioneer of outstanding importance.[14] A student of the Vienna Polytechnische Institut, from which he graduated in 1829, he taught applied mathematics and machine drawing from 1833 to 1841 at the Höhere Industrieschule in Zurich while at the same time working for Escher & Wyss, the Swiss mechanical engineering firm specializing in the production of water turbines. Throughout his career, his practical experience permeated his work as a teacher. The courses he taught at the Polytechnische Schule of Karlsruhe, as professor of mechanical engineering from 1841 until his death in 1863, and his books—notably a treatise on water turbines and a widely used textbook, *Resultate für den Maschinenbau*, both published in the 1840s—were exemplary in demonstrating how theoretical insights of the kind that had been strongly developed by the French engineers of the late eighteenth

13. Gustave-Adolphe Hirn, *Recherches sur l'equivalent mécanique de la chaleur* (Colmar, 1858), 20–43, 75–78, and Table D. Although Hirn's experiments confirmed the existence of a relationship between heat and work, variations in his values for J cast doubt on the existence of a universal mechanical equivalent of heat that would be constant in all circumstances. Soon afterwards, however, Hirn argued for the existence of an invariable mechanical equivalent of heat; see his "Confirmation expérimentale de la seconde proposition de la théorie mécanique de la chaleur...Démonstration analytique de cette proposition et conséquences principales auxquelles elle conduit," *Cosmos*, 22, 1er semestre (1863), 413–468.

14. On Redtenbacher, see Ferdinand Redtenbacher, *Biographische Skizze und Festbericht. Erinnerungsschrift zur siebzigjährigen Geburtstagfeier F. Redtenbacher's* (Munich, 1879), and Karl Lindner, *Ferdinand Jakob Redtenbacher, der Begründer des wissenschaftlichen Maschinenbaus. Vortrag, gehalten im Außeninstitut der Technischen Hochschule Graz, am 10. November 1959* (Graz, n.d.).

and early nineteenth centuries could inform and be informed by the realities of practice.[15]

Although the unified approach that Redtenbacher advocated took root slowly, attitudes similar to his were eventually assimilated everywhere. Even in Britain, William Rankine's teaching in the Regius chair of civil engineering and mechanics at Glasgow and his frequently reprinted manuals, first published in the late 1850s and 1860s, offered a powerful challenge to entrenched rule-of-thumb approaches.[16] By the 1870s, however, the lead in Europe had been taken by the Polytechnische Schulen and Technische Hochschulen in the German and Austro-Hungarian Empires. Here, a new emphasis on school-based instruction, conveyed in the lecture room, in the wave of textbooks of mechanical engineering that followed in the wake of Redtenbacher's and Rankine's publications, and in tentative steps to develop laboratory teaching, was reinforced by the much broader campaign to raise the status of higher technical education.

The process by which laboratory courses were introduced as an integral part of the curriculum provide a valuable touchstone for the changes that were afoot in mechanical engineering. As in the case of physics, such early precedents as there were had been rather distant ones. By the mid-century, a number of the more important technical schools and engineering departments did already possess laboratories, most of them devoted to mechanics. In this earlier period, however, the main purpose of such laboratories had been to support lecture demonstrations and the work of testing and calibration that professors were increasingly called upon to conduct for external clients. Even the pedagogical activities did not involve students, who, just as in physics departments, remained passive observers of the demonstrations that the professors and their assistants conducted for them. The similarity between engineering and physics was also close in that, like their colleagues in physics, teachers of engineering received little institutional support for experimental research. In fact, the cost of facilities for this purpose placed them beyond the reach of most teaching institutions, and, as we indicated earlier, it was only by acting as consultants for state-owned or private establishments that academic engineers could provide

15. Ferdinand Jakob Redtenbacher, *Theorie und Bau der Turbine und Ventilatoren* (Mannheim, 1844); almost as influential was his *Theorie und Bau der Wasserräder* (Mannheim, 1846). His *Resultate für den Maschinenbau*, containing diagrams, formulae, and rules for the guidance of mechanical engineers, went through four German editions between 1848 and 1860 and was translated into French. Although Redtenbacher displayed his respect for the theoretical work of Navier, Coriolis, Poncelet, and Morin, he felt that the gulf between theory and practice in the French tradition had always been, and still remained, wide.

16. See the references in Chapter 1 (ref. 69).

adequately for their experimental work. One consequence of this practice was that the work was conducted primarily outside the schools and universities and that students were, by virtue of this physical detachment, effectively excluded. Nevertheless, countervailing tendencies, in the emerging recognition that techniques assimilated in the laboratory were an essential ingredient in the training of a mechanical engineer, meant that by the 1870s the necessary facilities were beginning, very unevenly, to make their appearance.

One of the institutions in which the gathering pace of the turn to laboratory instruction and laboratory-based research in mechanical engineering was first manifested was the Polytechnische Schule in Munich, which took a conspicuous initiative soon after its foundation in 1868. Here, a main catalyst was the professor of mechanical engineering (*Maschinenlehre*), Carl von Linde. Still in his early thirties, von Linde applied for and obtained a grant from the Bavarian government which, in 1875, allowed him to create the school's, and Germany's, first mechanical teaching laboratory.[17] He used its facilities, which included two steam engines and associated equipment, as the base for his course until, three years later, he left what in 1877 had become the Technische Hochschule of Munich to devote himself to his private refrigeration company. Over the next decade, other Technische Hochschulen followed suit. By 1886, Dresden, Darmstadt, and Stuttgart all possessed substantial mechanical engineering laboratories, with facilities for student exercises in measurement (in particular, related to the strength of materials) and the study of the performance of steam engines,[18] and by 1888 Alois Riedler had established a base, in the chair of mechanical engineering at the Technische Hochschule of Berlin, for an ultimately very effective campaign in favor of research and teaching laboratories and against the excessively mathematical character of the training of mechanical engineers.[19]

17. On Linde's laboratory, see Hans Liudger Dienel, "Der Münchner Weg im Theorie-Praxis-Streit um die Emanzipation des Wissenschaftlichen Maschinenbaus," in Ulrich Wengenroth, ed., *Die Technische Universität München* (Munich, 1993), 87–115, on 93–94.

18. Excellent general sources on these and other developments in the Technische Hochschulen are Wilhelm Lexis, ed., *Die Technischen Hochschulen im Deutschen Reich* [vol. 4, part 1 of Lexis's *Die Unterrichtswesen im Deutschen Reich*] (Berlin, 1904), and *Die Deutschen Technischen Hochschulen. Ihr Gründung und geschichtliche Entwicklung* (Munich, 1941), a volume of historical sketches on the eighteen Technische Hochschulen that existed at the time in Germany and Austria.

19. Karl-Heinz Manegold, "Alois Riedler," in Wilhelm Treue and Wolfgang König, eds., *Berlinische Lebensbilder. Techniker* [Berlinische Lebensbilder, ed. Wolfgang Ribbe, vol. 6] (Berlin, 1990), 293–307.

Change in the functions of the laboratories came about slowly, and initially the main emphasis in the work carried out in them remained on routine commissioned work for local industries, professorial lectures and demonstrations, and, to some extent, the professors' personal research. In the 1870s, the Mechanisch-technische Laboratorium of the Technische Hochschule of Munich (directed by Johann Bauschinger) and the Königliche Mechanisch-Technische Versuchsanstalt of the Technische Hochschule of Berlin were typical in being largely devoted to testing; the same was true of the Materialprüfungsanstalt of the Polytechnikum of Stuttgart, although here as in the corresponding section of the Polytechnikum of Dresden, laboratory apparatus was also used for demonstrations to students.[20] Through the 1880s and more obviously in the 1890s, however, the "hands-on" instruction of students in experimental techniques came to be recognized, along with research, as a central activity. With this recognition, a new phase in the reorganization of the higher technical schools began, making well-equipped teaching laboratories an increasingly prominent feature. Gradually, such laboratories became symbols of the disciplinary autonomy of mechanical engineering and its final emancipation from the bonds that, until recently, had tied the study of machines either to rough and ready workshop practice or to the highly mathematical world of abstract theoretical mechanics.[21]

Although the ideal of a laboratory-based pedagogy was first properly realized in Germany, it soon had material repercussions in other countries as well. One of the earliest, and most admired, initiatives outside Germany was that of Alexander Kennedy in England. When, in 1875, Kennedy was appointed professor of engineering at University College, London, he

20. Alexander B.W. Kennedy, "The use and equipment of engineering laboratories," *Minutes of proceedings of the Institution of Civil Engineers, 88* (1886–1887), 1–80, on 73–76. On the Königliche mechanisch-technische Versuchsanstalt and the developments that by 1904 had given it a far broader range of responsibilities as the Königliche Materialsprüfungsamt, see Adolf Martens and Max Guth, *Königliche Materialprüfungsamt der Technischen Hochschule Berlin. Denkschrift zur Eröffnung* (Berlin, 1904). The changing emphases to which we refer in this paragraph are brought out clearly in the account of the Materialprüfungsanstalt at Stuttgart (first conceived in 1878) and the associated mechanical engineering laboratory in Gerhard Zweckbronner, "Die Ingenieurwissenschaften im 19. Jahrhundert," in Johannes H. Voigt, ed., *Festschrift zum 150 jährigen Bestehen der universität Stuttgart* (Stuttgart, 1979), 189-222, on 210-214.

21. See Victor-Auguste Dwelshauvers-Dery and Julien Weiler, *Referendum des ingénieurs. Enquête sur l'enseignement de la mécanique* (Liège, 1893), "Les laboratoires de mécanique et les écoles techniques supérieures," 3–62.

used his introductory lecture for the new academic year to state explicitly his program for a new kind of instruction.[22] Having been apprenticed for more than five years to a marine engineering firm before working for seven years as a consulting engineer, he had a keen sense of the changes that were occurring in engineering practice. He still firmly believed that pupillage on the job was an essential means of obtaining basic practical knowledge and skills. He was equally convinced, however, that the preparation of an engineer should also include practical training at a more advanced level. As he insisted, training of this kind could only be conveyed effectively in an institution of higher education, the factory being unsuitable since the immediate interests of the manufacturer would all too easily intrude. Also and more specifically, the training had to be given in engineering laboratories, fitted with machinery that would allow students to be drilled in the techniques of measurement and the application of rigorous approaches to the analysis of technical problems. The aim was to instill habits of accuracy and a familiarity with quantitative methods and experimental procedures, as applied to such areas as the performance of heat engines, the effects of friction, and the physical properties of materials and structures.[23] Within three years of Kennedy's arrival at University College, the scheme was implemented with the opening of a spacious basement laboratory, the first for students of mechanical engineering in England. Among the equipment was a steam engine for trial runs, an experimental boiler, testing machines for the measurement of the mechanical properties of materials, and a range of brakes, dynamometers, and other devices for studying performance and efficiency.[24]

The laboratories that William Cawthorne Unwin established in London, first at the Royal Indian Engineering College in Cooper's Hill in 1883 and, a year later, at the new Central Institution in South Kensington, closely resembled Kennedy's. For Unwin, it was a matter of principle that a mechanical laboratory and a workshop should exist in close proximity to each other and that both should be given equal prominence in any complete program. Workshop training took place in the first and second years, while

22. Alexander B.W. Kennedy, "The place of colleges in engineering training" [Introductory address delivered to the Faculties of Arts and Laws and of Science at University College, 5 Oct 1875], *Journal of the Society of Arts, 23* (1874–1875), 963–968, and *Engineering, 20* (1875), 284–285 and 315–316.
23. Kennedy, "The place of colleges in engineering training" (ibid.), in *Journal of the Society of Arts, 23* (1874–1875), 966–967, and *Engineering,* (ibid.), 316.
24. Kennedy, "The use and equipment of engineering laboratories" (ref. 20), 1–6. The laboratory measured 90 feet by 44 feet, and was 16 feet 6 inches high. A boiler-house and a large classroom for drawing, measuring 71 feet by 51 feet, were housed elsewhere in the building.

instruction in the laboratory was for students in their second and third years.[25] The second-year laboratory course offered "systematic instruction in every kind of testing of metals, stone, cement, &c.," as a means of obtaining "the best insight into the properties of materials and the laws of the strength of materials." In the third and final year, there was a program of "systematic experiments," to be conducted on a steam engine specially constructed for thermodynamic research. These experiments allowed the parameters affecting the economy and performance of the engine to be determined, the tests always being conducted, as the Central Institution's calendar proudly announced, "on a large and practical scale."

The number of mechanical engineering laboratories that were inaugurated in British university colleges during the 1880s (see Table 3.1) indicates how quickly what had been, in the previous decade, a rarity became a normal facility to be expected in virtually any school or department of engineering. The process was a costly one, and it was only made possible by the injection of substantial private funding attracted by the air of modernity that the laboratories in London and in many provincial institutions conveyed. In other industrialized countries, the literature emanating from engineering schools conveyed an equally clear vision of the value of laboratory teaching, although the fruits of the vision materialized at an uneven pace and often very slowly.

At the Ecole des Mines in Paris, for example, a much-needed new laboratory for mechanical engineering was opened only in 1902, though with the usual flourish that stressed its value both for student exercises and for research, notably Charles-Louis Frémont's work on the mechanical properties of metals.[26] Elsewhere, too, the quest for facilities yielded protracted frustration. One case that illustrates the capacity of local circumstances to impede the best-laid plans was that of Victor Dwelshauvers-Dery, a professor of applied mechanics at the Ecole Spéciale des Arts et Manufactures et des Mines and later at the Faculty of Science in the University of Liège and a tireless campaigner for investment in laboratories for his subject.[27]

25. The program is conveniently summarized in *City and guilds of London Institute. Programme. Technical education* (London, n.d. [c.1884]), 11–12.
26. E. Sauvage, "Le laboratoire de mécanique de l'Ecole Nationale Supérieure des Mines," *Annales des mines*, 10th ser., 7 (1905), 371–379.
27. On Dwelshauvers-Dery, see Armand du Chesne's biographical notice on him in Léon Halkin, ed., *Liber memorialis. L'Université de Liège de 1867 à 1935. Notices biographiques* (3 vols., Liège, 1936), 2, 434–455. Dwelshauvers-Dery's case and the evidence he had marshalled from correspondence with colleagues in several countries were published in an important series of articles, "Les laboratoires de mécanique et les écoles techniques supérieures," in *Le génie civil, 18* (1890–1891), 268–272, 285–287, 303–305.

Table 3.1
Mechanical engineering laboratories established in English institutions of higher education

Date of foundation	Institution	First Professor
1878	University College, London	Alexander B.W. Kennedy
1882	Mason Science College, Birmingham	Robert H. Smith
1883	Royal Indian Engineering College, Cooper's Hill, London	William C. Unwin
1883	University College, Bristol	Henry S. Hele-Shaw
1884	City & Guilds of London Central Institute, London	William C. Unwin
1885	Firth College, Sheffield	W.H. Greenwood
1886	Yorkshire College, Leeds	Archibald Barr
1888	Owens College, Manchester	Osborne Reynolds
1889	University College, Liverpool	Henry S. Hele-Shaw

Sources: Various, but especially Alexander B.W. Kennedy, "The use and equipment of engineering laboratories," *Min… proceedings of the Institution of Civil Engineers, 88* (1886-1887), 1-80, on 70-72.

Since 1869, Dwelshauvers-Dery had pursued a detailed study of steam-engine governors, in the face of difficulties that were only partially alleviated, from 1885, by the facilities made available to him by his friend Charles Beer, an engine manufacturer, at Jemmappes, in the mining area near Mons. Both his own research and his experience as a teacher convinced Dwelshauvers-Dery of the gap between even the most realistic treatment of the theory of steam engines and their actual performance. This led him to advocate what at the time would have been a highly original, even unique way of giving students immediate "hands-on" experience of such problems. As early as 1870, he approached the governing body of the Ecole Spéciale des Arts et Manufactures et des Mines to set up a teaching laboratory in which the most important facility would be a steam engine reserved exclusively for experimental use: the aim, in the words of Adolphe Hirn (on whose working engine at Le Logelbach, near Colmar, Dwelshauvers-Dery had taken readings in the 1870s), was to establish "a laboratory of industrial physics and mechanics in which young people can study in full-scale working conditions [*en grand et sur le vif*] and arrive, while studying, at useful results."[28] However, the governing body was unimpressed by what it evidently regarded as a strange aspiration for a department of mechanical engineering, and it was not until 1890, after repeated unsuccessful requests, that Dwelshauvers-Dery achieved his objective, in the context of a major reorganization that saw the Ecole Spéciale integrated as the dominant element in a new Faculté Technique.

Dwelshauvers-Dery was also responsible for an initiative that indicates the prominence of the issue among the teachers in higher technical schools. In 1890 he sent a questionnaire to colleagues and practicing engineers throughout Europe asking for their views on a variety of themes relating to the instruction of mechanical engineers, among them the state of the provision for "experimental teaching."[29] The results, which reflected the strong feelings of teachers, were widely circulated and read, not least in those countries where provision was still inadequate. One such country was Italy. Here, laboratories had been a central item in the agenda of the Museo Industriale of Turin at the time of its opening in 1862 in what was at the time the capital of Italy.[30] When, in 1866, the Museo was authorized to launch higher technical courses in industrial subjects, and when, one year

28. *Le génie civil, 18* (1890–1891), 304; also on p. 444 of the biographical notice cited in the previous note.
29. The resulting report is cited above in ref. 21.
30 *Il Regio Museo Industriale Italiano in Torino. 1862–1902* (Turin, 1902), esp. 5–17, and Anna Guagnini, "Higher education and the engineering profession in Italy: The *Scuole* of Milan and Turin, 1859–1914," *Minerva, 26* (1988), 512–548.

later, the courses became the core of the industrial engineering section of the Scuola di Applicazione per Ingegneri of Turin, a distinctive feature of the plan was the emphasis on laboratory-based instruction. Although a grant of £450 was earmarked for the fitting out of four laboratories (for technical physics, industrial chemistry and metallurgy, agricultural chemistry, and industrial mechanics), the new teaching program encountered numerous obstacles and underwent significant modification before it gained recognition. The difficulties meant that only two laboratories (technical physics and industrial chemistry) opened in 1869 and that the mechanics laboratory was not established until 1880. Moreover, and more importantly, through the 1870s, in the course of protracted negotiations with the Scuola di Applicazione, the plan for the laboratories lost the centrality it had had in the initial discussions. As a result, the prospects of obtaining the financial support that was necessary in order to equip them adequately became more remote; by 1880, as the comments of teachers make clear, the apparatus was being used, at best, for the traditional purposes of demonstration and a certain amount of testing commissioned by public and private customers.[31]

The history of the Museo Industriale shows how difficult it was to preserve the initial, highly original commitment to laboratory instruction, in the face of the quite different style of teaching that prevailed in the most advanced sector of technical education, that of the university-level *scuole di applicazione per ingegneri*. In the Scuole di Applicazione of Turin, in particular, instruction was essentially theoretical: workshop practice was not regarded as part of the five-year program, and even the equipment available for demonstrations was barely adequate for the purpose. The forces of tradition were equally tenacious at the most industrially oriented of the *scuole di applicazione*, the Istituto Tecnico Superiore of Milan. Here too, whatever equipment the school managed to purchase (always with difficulty and usually with the aid of private donations) was used only for demonstration and testing. It is true that the importance of the example set by the opening of the first teaching laboratories in other countries—those of the ETH were the favored model in Lombardy—was quickly recognized by some of the Milanese school's most far-sighted teachers, among them Giuseppe Colombo, professor of industrial mechanics from 1865 and director from 1897 to 1921.[32] Yet it was not until 1893 that a laboratory of

31. Guagnini, "Higher education and the engineering profession in Italy" (ibid.), 533-536.

32. Colombo's appointment in 1865 was as professor of "meccanica industriale e costruzione di macchine," although the title of his chair was subsequently changed on more than one occasion.

applied mechanics—widely regarded as a model for similar initiatives in other Italian schools—was created.[33]

Whenever and wherever they were introduced, and whatever form they took, the mechanical engineering laboratories that began to proliferate across Europe in the last quarter of the century enhanced the academic prestige of the departments that housed them, helping to make such departments more acceptable in the status-conscious world of higher education. The discussion that followed a paper that Alexander Kennedy gave before the Institution of Civil Engineers in London in 1887 is revealing. The published report on the discussion summarized the experiences of Osborne Reynolds as professor of engineering at Owens College, Manchester, where the introduction of a laboratory had provoked a sudden rise in the standing of the discipline. In the words of the report:[34]

> Since 1872 or 1873 [Reynolds] had been endeavouring to induce the authorities of the Owens College to increase the facilities of illustration and work for the engineering students; but although he had done his best to bring it about he had not succeeded until after the Author [i.e. Kennedy] had set the example, and he believed that the term "laboratory" had as much as anything conduced to his success. There was a great shyness about establishing a workshop, but the moment a laboratory was talked of, all opposition disappeared.

The almost magical power of the word *laboratory* is illustrated by the even more conspicuous instance of James Stuart's workshop in Cambridge. Three years after his appointment to the newly established professorship of mechanism and applied mechanics in 1875, Stuart obtained permission to set up a workshop and proceeded to equip it, largely at his own expense.[35] By 1881, he had secured enough support to induce the university to provide a new building for a gas engine and other machinery and machine tools. However, the presence of such a workshop on the Free School Lane site, an area designated for science at the very heart of the university, alarmed the representatives of the old order, who saw the emphasis on manual skills as an incongruous intrusion. It must be said that their concern was reinforced by Stuart's radical political views, which extended to his support for the rights of women. But the character and location of his

33. "I laboratori di meccanica applicata nel R. Istituto Tecnico Superiore di Milano," *Il Politecnico*, *44* (1896), 624–639.

34. "Discussion" following Kennedy, "The use and equipment of engineering laboratories" (ref. 20), 83.

35. Thomas J. N. Hilken, *Engineering at Cambridge University, 1783–1965* (Cambridge, 1967), 66–69.

workshop and the cost of maintaining it did more than anything else to make Stuart's position difficult and to provoke the resignation that he offered in December 1889 and put into effect in the following summer.[36]

Despite the conflict, the principle that engineering should be studied at Cambridge was not abandoned. Indeed, strong support for the discipline was voiced in the report of the university's specially appointed Mechanical Workshops Enquiry Syndicate, on which Osborne Reynolds, John Hopkinson, and Richard Glazebrook (the assistant director of the Cavendish Laboratory and future director of the National Physical Laboratory) were all leading figures,[37] and in fly-sheets and memoranda by these and other distinguished engineers and academics, including Alexander Kennedy and Horace Darwin (the youngest son of Charles Darwin and head of the Cambridge Scientific Instrument Company, founded in 1881).[38] The consensus to which the statements converged was that while an engineering course in Cambridge was highly desirable, it should only be mounted if its intellectual level was compatible with the general standing of the university. The solution lay in the creation of a completely new tripos whose title would incorporate the all-important word "science," complemented—significantly—by a well-founded, modern laboratory similar to Kennedy's in London. These steps reconciled the ideals of high scientific standards and traditional academic values with the pragmatic modernity of a technical

36. James Stuart to the Vice-Chancellor, 11 Dec 1889, published (with other relevant correspondence) in the *Cambridge University reporter, 1889–1890,* no. 821, 11 Feb 1890, 422.

37. The syndicate was established on 5 Dec 1889, less than a week before Stuart wrote his letter of resignation. See *Cambridge University reporter 1889–1890,* no. 811, 10 Dec 1889, 263.

38. The report and associated material were published in the *Cambridge University reporter 1889–1890,* no. 839, 4 Jun 1890, 822–845. Darwin, a staunch and very effective advocate of the creation of an engineering school at Cambridge, was well aware of the reasons behind the reluctance of the university authorities to approve the scheme. The emphasis he laid on the experimental side of engineering and on the centrality of work in a laboratory is reminiscent of Reynolds's comment quoted above. See, for example, the following passage in a fly-sheet by Darwin dated 2 Jun 1890:

> The present system of Engineering education has met with opposition from many members of the Senate....One possible modification would be to transform the Workshop into an Engineering Laboratory; this I think could be done at no great cost, and would remove many of the objections that have been raised.

A copy of the fly-sheet is tipped into the issue of the *Reporter* that John Willis Clark bequeathed to the Library of the University of Cambridge; see Cambridge University Library, Cam.b.1.10.20, immediately before issue no. 839, cited earlier in this note.

subject, and defused the hostility that existed in more conventional Cambridge minds. With the essential victory won, James Alfred Ewing, a relatively seasoned hand after twelve years in chairs of engineering, first at the University of Tokyo and then at the new University College of Dundee, arrived to succeed Stuart, and in the autumn of 1891, during his first year in Cambridge, the new approach was given material expression by the opening of the University's first engineering laboratory in buildings formerly occupied by the Perse School.[39] The first undergraduates to present themselves for the new Mechanical Sciences Tripos did so in the spring of 1894.

In addition to promoting the status of engineering within higher education, laboratories fulfilled the no less important function of fashioning the image that academic engineers presented to the outside world, in particular to the members of the engineering profession and the potential entrants to their courses. At a time when industry and practicing engineers were still profoundly marked by the traditions of on-site, shop-floor experience, laboratory exercises and instruction in the practice of measurement and standardization allowed institutions of higher technical education at last to identify a distinctive contribution which they were in a position to make and which the older forms of training, via apprenticeship and practical instruction on the job, could not provide. The strategy worked, and laboratories not only multiplied, as Table 3.1 indicates with respect to Britain, but were also quickly assimilated as a staple element in the public face that institutions fashioned for themselves in their relentless endeavors to win students. It is no coincidence that about this time calendars and prospectuses began to carry alluring photographs that flaunted laboratories as marks of excellence and proper targets for future benefactions, most of which came from local engineering and manufacturing companies.[40]

By 1900, mechanical engineering had come to be respected and well provided for in all countries that possessed even a modest system of advanced technical education: it had students (even if they were not always specializing in the subject), a recognized *cursus* that combined theory with a new perception of what constituted appropriate "practice" for a young engineer, and a status that would have been inconceivable half a century before. Until the later stages of this development, the discipline enjoyed a unique position as the quintessential embodiment of university-level engi-

39. Hilken, *Engineering at Cambridge University* (ref. 35), 112–118.
40. See, for example, the description of the engineering laboratory in the *Calendar* of University College, Bristol, for 1896–1897, 78–80, and the photograph of the experimental steam engine and dynamos on p. 52. Cf. also the glowing description of the newly opened engineering laboratory in the Preface to the *Calendar* of University College, Liverpool, for 1900–1901, xvi.

neering. But the entry of electricity into the pantheon of engineering specialties meant that from the 1890s mechanical engineering began to share its status with a partner that immediately assumed an at least comparable academic dignity. Electrical engineering profited conspicuously from the standing that it had taken mechanical engineers so long to achieve: to its great advantage, it easily appropriated many of the usages of mechanical engineering, in particular the emphasis on laboratory instruction.

The creation of laboratories of electrical technology

Before the 1880s, electricity (mainly with an orientation to telegraphy) tended to be taught as a minor part of the curricula in mechanical or civil engineering, usually under the broad umbrella of what was described as technical or applied physics. From the early 1880s, however, the first dedicated programs in the subject were created. These programs, which attracted large numbers of students and encapsulated the excitement of a discipline at the cutting edge of scientific and technological progress, squared well with the already rapidly changing face of higher technical education. The central place they accorded, from the start, to teaching laboratories was one way in which they did this: in contrast with mechanical engineering, workshops were generally not part of the pedagogical culture of electrical engineering, even in the earliest years of the discipline.

The assimilation of electrical technology into the institutions in which it found a home was exceedingly rapid. Institutions were eager to welcome a branch of engineering that elegantly reinforced the image they were seeking to present to the world; they also perceived the benefits to be gained from the public infatuation with electricity that made its mark everywhere in the wake of the electrical exhibition in Paris in 1881. However, the way in which the assimilation occurred differed significantly, at least initially, from one country to another. Broadly, there were two patterns. One approach favored the introduction of electrical courses as elements within the curriculum offered to mechanical engineers. The other was based on the creation of advanced programs, entirely devoted to electrical technology and usually lasting one year, that were designed for students who had already completed their studies in mechanical or civil engineering. These two approaches coexisted through the 1880s, and it was only in the 1890s that they were joined by the different and now more familiar pattern of self-contained, dedicated programs in what was variously described as electrotechnics (*Elektrotechnik*, *électrotechnique*, and *elettrotecnica* being the favored terms in Germany, France, and Italy), electrical technology, electrical engineering, or applied electricity.

These later programs, typically lasting three years and leading to specialized degrees and diplomas, are treated later in the section entitled, "Expansion and diversification." First, however, we discuss the two dominant patterns of the 1880s, which bear centrally on the main themes of this book, partly because of the prominence they both (though especially the first of them) gave to laboratory-based instruction, and partly because of their facilitating role in the emergence of electrical engineering as a pacemaker in moves to enhance the profile of research as a recognized activity in higher technical education.

Early developments: The first pattern

The first of the two patterns we have identified as co-existing in the 1880s was most obviously, though not exclusively, characteristic of courses instituted in Germany and Switzerland.[41] In both countries, the association with mechanical engineering remained close throughout the decade, allowing electrical technology to benefit from the prominence that mechanical engineers had acquired in the Polytechnische Schulen and Technische Hochschulen over the preceding quarter of a century. The discipline also benefited from developments outside the educational system. The most visible of these came from the German and Swiss electrical construction industry. Companies such as Siemens & Halske in Germany and Oerlikon in Switzerland moved promptly into the new field opened up by Edison's innovations in lighting. A main problem arising from the shift from the practices of traditional arc-lighting to those of incandescent lighting was the need for more powerful generators, which in turn created a demand for engineers whose training combined mechanical know-how with a specialism in the construction and installation of large-scale machinery and a command of the latest developments in electrical technology. A public plea by Werner von Siemens, who in December 1881 urged the creation of a chair of *Elektrotechnik* in every Technische Hochschule, reflected the concern and expressed the mood of the more progressive manufacturers.[42] No voice could have carried greater weight or done more to pave the way for the creation, in the following year, of the first institute and chair of electrical technology in the world at the Technische Hochschule of Darmstadt.[43]

The first holder of the chair at Darmstadt was the thirty-year-old Erasmus Kittler, fresh from a doctorate in physics, completed under Friedrich

41. On these courses, as they existed in Germany, see the excellent study of the establishment of the discipline of *Elektrotechnik* in Wolfgang König, *Technikwissenschaften. Die Entstehung der Elektrotechnik aus Industrie und Wissenschaft zwischen 1880 und 1914* (Chur, 1995).

42. Ibid., 87–96.

43. However, as Wolfgang König has observed, Siemens's support for the emergence of electrical technology as an independent academic speciality was with-

Kohlrausch at the University of Würzburg,[44] and a short period of teaching as a *Privatdozent* at the Technische Hochschule of Munich, where he also played a minor but active role in the international electrical exhibition of 1882.[45] With support from a variety of sources, including the city of Darmstadt, Kittler installed three laboratories, in which students were provided with what he regarded as their core training in the experimental study of electrical machinery, as well as space and equipment for photometric measurements. In the emphasis he placed on laboratory exercises, Kittler remained close to the most advanced practices in contemporary mechanical engineering, and it is no coincidence that he resolutely sustained the bonds between *Elektrotechnik* and *Maschinenbau*. Although the decision of 1883 that *Elektrotechnik* should become a separate, sixth section at Darmstadt recognized the distinctiveness of the discipline, it did so, significantly, without undermining the principle that mechanical and electrical engineers should establish common roots by following the same program in the first two years of the four-year course.

As we have stated, from the 1890s, engineering schools were to opt for a greater degree of separation between the electrical and mechanical interests to whose unity Kittler attached such importance. But the institutional emancipation of electrical technology was pursued with caution. Indeed, some of the largest and most successful institutions in continental Europe maintained a largely unified approach until after the turn of the century, by which time separate complete programs in electrical engineering had become far more common. At the Technische Hochschule of Berlin, for example, the facilities for *Elektrotechnik* were outstanding. Yet the subject remained an undifferentiated speciality within the section of *Maschinen-Ingenieurwesen* as late as 1928 and only became a separate faculty in 1955.[46]

At the Eidgenössische Technische Hochschule (ETH) in Zurich, too, mechanical and electrical engineering were taught for many years in close

---

drawn in 1884, when he took a conspicuously jaundiced view of the growth of the discipline. See ibid., 91–96.

44. Kittler's doctoral thesis, presented in 1880, was entitled "Spannungsdifferenzen zwischen sich berührenden Flüssigkeiten." See ibid., 16.

45. For accounts of Kittler's work, see ibid., esp. 15–20, and Wolfgang König, "Erasmus Kittler und die Frühzeit der Elektrotechnik," xv–xlii, in the reprint of Kittler's *Handbuch der Elektrotechnik in zwei Bänden* (1886–1890), published in 1986 by the VDI Verlag, Düsseldorf.

46. Reinhard Rürup, "Die Technische Universität Berlin 1879–1979: Grundzüge und Probleme ihrer Geschichte," in Reinhard Rürup, ed., *Wissenschaft und Gesellschaft. Beiträge zur Geschichte der Technischen Universität Berlin 1879–1979* (2 vols., Berlin, Heidelberg, and New York, 1979), *1*, 3–47, on 34; also the chronological appendix by Renate Schröder-Werle, ibid., *1*, 568 and 575.

FIG 15. The main building of the Berlin Technische Hochschule at Charlottenburg, c. 1890. From *Die Königliche Technische Hochschule Berlin — Charlottenburg* (Berlin, 1885), plate 1. Reproduced with permission of the Plansammlung, Universitätsbibliothek, Technische Universität, Berlin.

FIG 16. The machine room of the electrotechnical laboratory of the Berlin Technische Hochschule at Charlottenburg, c.1900. From Wilhelm Lexis, *Die Technischen Hochschulen im Deutschen Reich* (Berlin, 1904), facing p. 137. Reproduced with permission of the Bodleian Library, University of Oxford (*shelfmark* 2603 d 37).

proximity to each other in a single "Mechanisch-Technische Abteilung."[47] This structure did not prevent the institution from becoming an acknowledged pacemaker in the training of electrical engineers—a status achieved in large measure through an emphasis on laboratory exercises. The ETH's position in the field developed by a process of cautious accretion in which an electrical program that initially, in 1882, consisted of little more than lectures on galvanic phenomena and the mathematical theory of electrical currents by the professor of mathematical and technical physics, Friedrich Weber, was gradually expanded to include, by 1900, a treatment of all the latest technologies relating to the generation, distribution, and application of alternating current. When, in 1890, an Institut für Elektrotechnik was fitted out as part of the ETH's department of physics, its four lavishly equipped laboratories were regarded as a model for emulation: the palatial new physics building in which the Institut was located, along with all the laboratories for both pure and applied physics, was a state-of-the-art construction that offered an exceptional degree of protection from the external influences that normally impaired the precision of electrical instruments.[48] Despite the sophisticated facilities, however, the unity of the training for mechanical and electrical engineers remained a cherished principle until 1911, when a new freedom of choice allowed students to specialize in one or the other of the branches in the third and fourth years. The arrival of Karl Kuhlmann, the head of the high-voltage section of AEG, to occupy the newly created chair of theoretical electrotechnics and the associated break with the dependence on electrical teaching provided from within physics took the process a stage further in 1912 (the year of Weber's death). But it was still another twenty-three years before *Elektrotechnik* became a section in its own right, fully independent of *Maschineningenieurwesen*.

Although the best-known instances of electrical engineering prospering in a context either within or very close to mechanical engineering are to be found in Germany and Switzerland, the model could and did take root elsewhere. In France, the earliest and most notable instance of the unified approach, and one that proved to be a tenacious survivor, occurred in the country's most prestigious school for the training of industrial engineers, the Ecole Centrale des Arts et Manufactures. Here, instruction was offered

47. On the slow separation of electrical from mechanical engineering at the ETH, see Ernst Dünner et al., "Die Abteilung für Elektrotechnik," in *Eidgenössische Technische Hochschule. 1855–1955. Ecole Polytechnique Fédérale* (Zurich, 1955), 424–443.

48. G. Lasius, "Die Gebäude der Eidgenössischen Polytechnischen Schule," in *Festschrift zur Feier des fünfzigjährigen Bestehens des Eidg. Polytechnikums* (2 vols., Zurich, 1905), 2, 336–343. The physics building was erected at a cost of £85,000 (£60,000 for the structure and land, and £25,000 for equipment).

in applied electricity from 1884, when a *maîtrise de conferences* in electricity—the first such post in France—was created.[49] But the discipline (taught by Démétrius Monnier, a former *centralien* who had spent most of his career as a gas engineer) remained dominated, as it always had been, by the syllabus into which it was injected and hence by the preponderant attention that was paid at Centrale to mechanical engineering. Even the opening of a specialized electrical laboratory in 1894 (for final-year pupils working on electrical measurements and motors) and of a general laboratory for student use six years later did not significantly change the balance, any more than the new electrical laboratories of the Ecole des Mines did at about the same time.[50] It is an equally telling reflection on the importance attached to mechanical engineering as an essential foundation for electrical engineers that the absence of courses on *mécanique appliquée* in the Ecole Municipale de Physique et de Chimie Industrielles—a progressive school whose creation by the city of Paris in 1882 coincided with the dawn of the new age of electricity—was seen as a grave impediment to the institution's endeavors to pursue the training of electrical engineers and hence as a handicap for the program.[51]

In England, academic engineering was also somewhat slow to embrace the new discipline, at least by the standards of Germany. One reason for this lies in the extent to which a number of the leading departments of physics, among them University College, London, Owens College in Manchester, and University College, Liverpool, had developed strong teaching and research interests in telegraphy and other aspects of applied electricity since the 1860s, making physics a natural disciplinary home for the new departures in lighting and power generation in the early 1880s. Nevertheless, from the mid-1880s, courses in electrical technology began slowly to emerge. The model in this respect, and a highly successful one, was the program offered by the City and Guilds Central Institution (from 1893 the Central Technical College) in South Kensington. Here, in purpose-built premises, financed by the City and Guilds of London Institute and under the control of William Ayrton, a degree course in electrical engineering

49. Léon Guillet, *Cent ans de vie de l'Ecole Centrale des Arts et Manufactures 1829–1929* (Paris, 1929), 122–123.
50. Ibid., 123–124. On the two laboratories for electrical measurements and the rooms for photometry and the testing of electrical machinery that were installed at the Ecole des Mines in 1902, see H. Roberjot, "Le laboratoire d'électricité de l'Ecole Nationale Supérieure des Mines," *Annales des mines*, 10th ser., "Mémoires," 7 (1905), 527–540. Roberjot also gives a brief account of the three-year program of practical exercises that students conducted in the new premises.
51. This view was expressed by Georges Claude, a former pupil of the school, in an editorial in *L'étincelle électrique*, 1 (1894–1895), 25–26.

FIG 17. The electrical laboratory of the City and Guilds Central Institution, London, 1888. According to Ayrton, the laboratories of the Central Institution lay at the heart of the teaching, "the knowledge of the experimental sciences being obtained more by the students themselves working in the physical and chemical laboratories attached to the school than from lectures or bookwork;" see *The Committee of the City and Guilds of London Institute. A technical school for London* (London, n.d. [1880], 4). Lectures were given, but in Ayrton's conception they "should be rather for the purpose of guiding the students' minds in their laboratory work rather than to give them mere book knowledge." From "L'enseignement de l'électricité industrielle en Angleterre," *La lumière électrique, 30* (1888), 459-465, on 460. Private collection.

FIG 18. The mechanical workshop and laboratory of the City and Guilds Central Institution, London, 1888. The syllabuses for first-year students in the departments of engineering and physics were almost identical. In both cases, classroom instruction in mechanics and instruction in the mechanical workshop and laboratory occupied more than a third of the weekly program. From "L'enseignement de l'électricité industrielle en Angleterre," *La lumière électrique*, 30 (1888), 357-362, on 361. Private collection.

was established in 1884 as one of the institution's three specialties, the others being in mechanical engineering and chemistry.[52]

From the start, Ayrton signaled his intention of striking out on a distinctive track that reflected both his belief in the central importance of laboratory instruction and the breadth of the interests he had developed since his days as a student of Carey Foster in the department of physics at University College in the 1860s.[53] In the six years for which he had served as professor of physics and telegraphy at the Imperial College of Engineering in Tokyo (1873–1879) and in his time as professor of applied physics in London at the City and Guilds Technical College, Finsbury (1879–1884), he had worked closely with mechanical engineers, including John Perry, who held chairs of mechanical engineering both in Tokyo and at Finsbury.[54] Now at the Central Institution he forged similar alliances, notably with William Unwin, the professor of engineering, while also maintaining a fruitful collaboration with Perry, who had stayed on at Finsbury. These associations, which he forcefully articulated in his public statements, made their mark on his pedagogical practices.[55] For while Ayrton maintained teaching laboratories for magnetism, heat, and sound, so conscientiously covering the whole range of subjects embraced by his chair, which was formally in physics, the main focus of his own interests and of those he communicated to his students was firmly in applied electricity. The level of enrollments fully

52. Between 1884 and 1888, a total of £6470 was spent on fittings and apparatus for the physical laboratories. See *City and guilds of London Institute. Report of the Board of Studies on the equipment of the Central Institution* (London, 1889), 9. Although Ayrton's department was initially designated a department of physics, the degree that it offered was in electrical engineering. By the mid-1890s, the reality of the situation had been acknowledged by the adoption of the title "Department of Physics and Electrical Engineering."

53. In the absence of a full-scale biography of Ayrton, see the obituary notices in *Journal of the Institution of Electrical Engineers, 43* (1908), 821–822; *Minutes of the proceedings of the Institution of Mechanical Engineers, 177* (1908), 295–296; and *Proceedings of the Royal Society of London*, ser. A, *85* (1911), "Obituary notices of Fellows deceased," i–viii (by "J.P.," presumably John Perry).

54. William E. Ayrton and John Perry, "Some remarks on the technical education of an electrical engineer," *Journal of the Society of Telegraph and Electrical Engineers, 11* (1882), 389–398. On the collaboration between Ayrton and Perry at Finsbury, see William H. Brock, "Building England's first technical college: The laboratories of Finsbury Technical College, 1878–1926," in Frank A.J.L. James, ed., *The development of the laboratory. Essays on the place of experiment in industrial civilisation* (Basingstoke and London, 1989), 155–170.

55. For an assessment of Ayrton's work, see Dominic W. Jordan, "The cry for useful knowledge: Education for a new Victorian technology," *Proceedings of the Institution*

endorsed the emphasis he chose and the appeal of the subject. His lectures and laboratory classes were better-attended than others anywhere in the institution, and the facilities supporting them continued to absorb more than their share of the initial endowment.

Early developments: The second pattern

In Belgium, Italy, and, to a lesser extent, France, a variety of circumstances, including the historical vested interests of the mechanical and civil engineering professions and, above all, the inflexibility of the structures of higher education, made the approach we have just described hard to implement. The broad, predominantly theoretical conception of what constituted the most advanced forms of engineering in these countries was at once too entrenched in curricula and too central to the high status that the most prestigious engineering qualifications enjoyed to allow the easy introduction of new specialties, especially ones rooted in untried, rapidly evolving technologies. With the core of the existing teaching programs virtually unassailable, innovators looked instead to "postgraduate" courses that could be offered to candidates who had already obtained their main engineering qualification via a more conventional *cursus*.

The pacemaker and main model for this approach was the Institut Montefiore in the Belgian industrial town of Liège. This highly original institution grew from the initiative of Georges Montefiore, a wealthy Jewish engineer and entrepreneur of British birth and (until his naturalization as a Belgian citizen in 1882 at the age of 49) British nationality. After graduating from the Ecole des Arts et Manufactures in Liège in 1852, Montefiore developed extensive interests in the production of nickel and other metals and in railways, and held shares in a number of companies in Italy and elsewhere in Europe.[56] A keen observer of technological innovations who had come back to Liège deeply impressed by what he had seen at the electrical exhibition in Paris in 1881, he made available £4000 with which to fit

of Electrical Engineers, 132 (1985), 587–601, and Graeme J.N. Gooday, "Teaching telegraphy and electrotechnics in a physics laboratory: William Ayrton and the creation of an academic space for electrical engineering in Britain," *History and technology, 13* (1991), 73–111.

56. On the Institut Montefiore and its founder, see *Université de Liège au Sart-Tilman. 100ᵉ anniversaire de l'Institut d'Electricité Montefiore* (Liège, 1938), and Philippe Tomsin, "L'Institut électrotechnique Montefiore à l'Université de Liège, des origines à la Second Guerre mondiale," in Laurence Badel, ed., *La naissance de l'ingénieur-électricien. Origines et développement des formations nationales électrotechniques* (Paris, 1997), 221–232. A convenient biographical sketch of Montefiore is in the Belgian *Biographie nationale, 38* [supplement, vol. 10] (Brussels, 1974), 595–618.

FIG 19. Testing room for dynamos and other machinery at the Institut Montefiore, Liège, 1886. The equipment in the room included two five-horse-power turbines, an Otto two-cylinder gas engine, six dynamos of various types (Edison, Siemens, Gramme, Weston, and in two cases Gramme-Jaspar, one of the last-named being used for laboratory demonstrations), and a number of dynamometers. Students were trained in this laboratory in the practices of industrial measurement. From *Institut Electro-technique Montefiore annexé à l'Université de Liège. But -- programme -- conditions d'admission* (Liège, 1886), plate facing p. 10. Reproduced with permission of the British Library (*shelfmark* 8757.bb.15(1)).

out a new school for the instruction of electrical engineers, on the understanding that the government would provide the building and pay the teaching staff. The school opened in 1883 and soon became one of the professional *écoles spéciales* of the University of Liège. It offered two programs. One, lasting two years, admitted (by examination) candidates who had passed through the preparatory school for entry to the Ecole Spéciale des Mécaniciens of Liège or attended schools at a similar level: in the first year of this program, students followed courses in applied mechanics, industrial physics, metallurgy, mechanical engineering, and foreign languages, and it was only in the second year that they devoted themselves to electrical engineering. The second program, which was in every sense the flagship of the institution, lasted one year and was available exclusively to those who already held degrees or diplomas at the highest level in engineering or physics.[57]

The core of the one-year course, which was devoted entirely to electricity and its applications, was a series of laboratory exercises occupying six hours each day. At the beginning of the course, students were introduced to the construction and manipulation of electrical instruments, with an emphasis that consistently reflected the changing patterns of technology. In the 1880s, time was devoted mainly to telegraphy and lighting. But when, in 1891, the school moved to new and roomier premises, with a laboratory refurbished through another, even more generous donation by Montefiore, more modern steam engines and motors were installed. This equipment, which increased the total power available four-fold, allowed students to carry out unprecedentedly elaborate experimental work and made it possible to introduce new areas of study, notably high-voltage generators and transformers and the application of electricity to metallurgy and traction.[58]

A key figure in the success of the Institut Montefiore was the school's first director and professor of "électrotechnique," Eric Gérard. A graduate of the Ecole des Mines in Liège who had begun his career in the late 1870s in the government's telegraphic service and studied for a year at the Ecole

57. *Institut Électro-Technique Montefiore annexé à l'Université de Liège. But–programme–conditions d'admission* (Liège, 1886). See also Giulio Bertolini, "L'enseignement de l'électrotechnique et l'Institut Electro-Technique Montefiore. Annexé à l'Université de Liège," *Bulletin de l'Association des Ingénieurs-électriciens sortis de l'Institut Electro-Technique Montefiore*, 2nd ser., *1* (1890), 60–77.

58. "Les nouvelles installations de l'Institut Electro-Technique Montefiore," *Bulletin de l'Association des Ingénieurs-électriciens sortis de l'Institut Electro-Technique Montefiore*, 2nd ser., *1* (1890), 344–352.

FIG 20. The precision measurement laboratory of the Institut Montefiore, Liège. From *Institut Electro-technique Montefiore annexé à l'Université de Liège. But—programme—conditions d'admission* (Liège, 1886), plate facing p. 10. Reproduced by permission of the British Library (*shelfmark* 8757.bb.15(1)).

FIG 21. The new premises of the Institut Montefiore, Liège, inaugurated in 1891. Reproduced with permission from the Collection du Centre d'Histoire des Sciences et des Techniques de l'Université de Liège.

Supérieure de Télégraphie in Paris, Gérard established an international reputation as a teacher, chiefly through his *Eléments d'électrotechnique* (1886) and his even more widely used textbook, *Leçons sur l'électricité* (1890).[59] Despite his youth—he was just 27 when he was appointed—his personal standing, allied to the high quality and relevance of the training that was offered and the relative opulence of the facilities, helped to make the Institut Montefiore an object of great interest to observers in other countries. One indicator was the substantial number of students who came from abroad: of the 201 who enrolled from the opening in 1883 to 1891, ninety were from outside Belgium.

The Institut Montefiore made its greatest mark internationally as a model for developments in Italy, which provided a large proportion of the school's foreign students. In both Milan and Turin one-year programs modeled ex-

59. Eric Gérard, *Eléments d'électrotechnique* (Liège, 1886), and *Leçons sur l'électricité* (2 vols., Paris and Liège, 1890; and six subsequent editions by 1914). On Gérard, see the biographical notice by Omer de Bast in Halkin, *Liber memorialis. L'Université de Liège de 1867 à* 1935 (ref. 27), 2, 476–478. In fact, Gérard's teaching career had begun in 1882, when he succeeded Frédéric Delarge in the chair, also in "électrotechnique," that had been established at the University of Liège in 1880.

FIG 22. A laboratory for electrical measurement in the new Institut Montefiore, c. 1910. Reproduced with permission from the Collection du Centre d'Histoire des Sciences et des Techniques de l'Université de Liège.

plicitly on those of Liège were in place by 1887.[60] In Milan the initiative came from Giuseppe Colombo.[61] It was Colombo who, as early as 1883 and while serving as professor of industrial mechanics at the Istituto Tecnico Superiore, organized the establishment of the first commercial electric power station in Italy, located in the heart of Milan, and then, in 1884, became the managing director of the Italian Società Edison, a subsidiary of the American Edison Electric Light Company that later assumed independence as a producer and distributor of electricity (the most important in the field in Italy). As part of his attempt to promote the electrical industry in Italy, Colombo introduced courses in electrical engineering into the general program of the Istituto Tecnico Superiore. In doing so, he recognized the constraints on his strategy. As he knew, the creation of a formally constituted separate section within the school would have entailed delays due partly to bureaucratic slowness but also to the reluctance of the Italian Ministry of Education and of the university authorities to promote specialization. To Colombo and his colleagues, such delays were unacceptable; as they saw it, the most expeditious way of forming a small but competent group of electrical experts was simply to graft the new instruction on to the existing pedagogical structure in the form of an additional year of specialized study.

The type of instruction that Colombo favored rested essentially on the use of instruments and machinery, and it was his insistence on this point that persuaded the board of the Istituto Tecnico Superiore to follow Liège in privileging laboratory-based teaching. In 1887 the Istituzione Elettrotecnica Carlo Erba, attached to the school but for all practical purposes independent, opened its doors, offering a one-year advanced course.[62] Like the Institut Montefiore, the new institution depended on funds provided by a private donor—in this case, a chemical manufacturer, Carlo Erba, who offered almost £16,000, the income from which was to pay for the equipment and maintenance of a new laboratory and the salaries of the teachers.

60. For a comparative account of the origins and early developments of these two initiatives, see Anna Guagnini, "The formation of the Italian electrical engineers: The teaching laboratories of the Politecnici of Turin and Milan, 1887–1914," in Fabienne Cardot, ed., *Un siècle d'électricité dans le monde. 1880–1980* (Paris, 1987), 238–299.

61. On Colombo, see Carlo G. Lacaita, "Giuseppe Colombo e le origini dell'Italia industriale," in Lacaita, ed., *Giuseppe Colombo. Industria e politica nella storia d'Italia. Scritti scelti: 1861–1916* (Bari, 1985), 5–86, and R. Cambria, "Giuseppe Colombo," in *Dizionario biografico degli Italiani, 27* (Rome, 1982), 213–228.

62. *L'Istituzione Elettrotecnica Carlo Erba annessa al R. Istituto Tecnico Superiore di Milano. 1887–1900* (Milan, 1901).

The influence of the Belgian institution loomed large in the choice of the first director and professor of electric measurement, Luigi Zunini, a former graduate of Montefiore and assistant to Eric Gérard. Yet there were important differences between the Belgian and Milanese initiatives. One lay in the limited financial resources that were available in Milan. For although Erba's donation was substantial, the regular income was rather modest. In fact, the laboratory's original equipment was largely restricted to instruments for measurement. By the end of the century, further private donations had allowed some refitting, but it was only in 1904 that the institution moved into a new purpose-built laboratory and only then that more powerful and updated d.c. and a.c. generators and motors were made available for the students' practical exercises.[63]

Another departure from the Montefiore model lay in the modest number of admissions. The maximum annual intake was 14 students, selected on merit from civil and mechanical engineering graduates and the best final-year students (again, in civil and mechanical engineering) of the Istituto Tecnico Superiore. Clearly, the aim of the Istituzione Elettrotecnica was to train a small technical elite that was expected to pave the way for the extension of the application of electricity in Italian industry, especially in Lombardy. In fashioning this technical elite, a sound general preparation, rather than originality, was the priority; as a result, although the emphasis was firmly on laboratory-based instruction, research remained a marginal activity. Certainly, by the end of the century, a small research laboratory had been equipped, and Riccardo Arnò, one of the members of the staff, worked there on measuring instruments, in particular on electricity meters and on the behavior of dielectrics in rotating electromagnetic fields.[64] However, students were not encouraged to engage in independent inquiries, nor indeed to take part in such research as their teachers might pursue. The laboratory facilities of the Istituzione Elettrotecnica, like those of the school to which it was attached, were to remain essentially, and successfully, devoted to teaching, complemented by testing (mainly for local manufacturers), until well into the twentieth century.[65]

63. *Nel cinquantenario della Istituzione Elettrotecnica Carlo Erba. 1887–1937* (Milan, 1937).

64. *L'Istituzione Elettrotecnica Carlo Erba* (ref. 62), esp. 18–23, and Piero Regoliosi and Andrea Silvestri, "L'Istituzione Elettrotecnica Carlo Erba," in Enrico Decleva, ed., *Il Politecnico di Milano nella storia italiana (1914–1963)* (Bari, 1988), 423–449.

65. The surveys of the occupations of former students published in *L'Istituzione Elettrotecnica Carlo Erba* (ref. 62), 27–33, and in *Nel cinquantenario della Istituzione Elettrotecnica Carlo Erba* (ref. 63), 89–122, suggest that the main objectives of the institution were realized. The majority of the students who graduated from the Istituzione Elettrotecnica, especially those who did so before 1905, quickly rose to

In Turin, too, implementation of the Montefiore model was the result of an independent decision taken by a local institution—in this case primarily to circumvent the rigid regulations of the Italian university system to which the engineering schools belonged. The mastermind was Galileo Ferraris, professor of technical physics at the Museo Industriale of Turin (the industrial engineering branch of the Scuola di Applicazione per Ingegneri, later to become the Politecnico of Turin).[66] Although he regarded himself first and foremost as a scholar and a teacher and directed his main research interests to the theoretical aspects of electrical engineering, Ferraris was active as an advisor to a number of public and private initiatives concerned with the industrial application of electricity. He was involved, for example, in the organization of the Esposizione Generale Italiana in Turin in 1884.[67] For this exhibition, he mounted a demonstration of alternating current transmission using Gaulard's transformer—a precursor, on a smaller and less ambitious scale, of the one at the Frankfurt exhibition in 1891—and went on to further investigations of alternating currents and rotating magnetic fields, culminating in one of the essays that laid the foundation for the development of the alternating current motor.[68]

In the wake of the Turin exhibition Ferraris, in 1886, proposed the creation of a teaching laboratory attached to the Museo Industriale. In the following year the Scuola e Laboratorio di Elettrotecnica was inaugurated.[69]

---

important positions in electricity-related branches of industry. This does not necessarily testify to the high standards of the courses but rather to the opportunities that graduates were able to exploit and to create for expansion in the industrial applications of electricity, notably though not exclusively in the local context. On the industrial contribution of graduates of both the Milanese and the Turinese schools, see Carlo G. Lacaita, "Politecnici, ingegneri e industria elettrica," in Giorgio Mori, ed., *Storia dell'industria elettrica in Italia*. Vol. 1. *Le origini, 1882–1914* (Rome and Bari, 1992), 603–644.

66 Guido Grassi, "Cenno sulle opere di Galileo Ferraris," published in *Opere di Galileo Ferraris* (3 vols., Milan, 1902–1904), *1*, vii–xxiii.

67. Ibid., xiii–xv. On the 1884 exhibition, see Edoardo Daneo, *Esposizione Generale Italiana in Torino 1884. Relazione generale* (2 vols., Turin, 1886).

68. Galileo Ferraris, "Rotazioni elettrodinamiche prodotte per mezzo di correnti alternate," *Atti della R. Accademia delle Scienze di Torino, 23* (1887–1888), 360–375; reprinted in *Opere di Galileo Ferraris* (ref. 65), *1*, 333–348.

69. *La Scuola con Laboratorio di Elettrotecnica "Galileo Ferraris" del Regio Museo Industriale Italiano in Torino* (Turin, 1899). A detailed study of this institution is Alessandra Ferraresi, "Nuove industrie, nuove discipline, nuovi laboratori: la Scuola superiore di elettrotecnica di Torino (1886–1914)," in Enrico Decleva, Carlo G. Lacaita, and Angelo Ventura, eds., *Innovazione e modernizzazione in Italia fra otto e novecento* (Milan, 1995), 376–494; see also Sigfrido Leschiutta and Anna Maria Rietto, "The formation of electrical engineers in Torino," in Badel, *La naissance de l'ingénieur-électricien* (ref. 56), 319–340.

Like the Institut Montefiore and the Istituzione Elettrotecnica in Milan, the Scuola of Turin accepted students who had already graduated, in this case in civil or mechanical engineering, mathematics, or physics, or from military schools. The criteria for admission were rather more relaxed than in Milan: as a result, enrollments grew from 26 in 1890 to as many as 128 in 1898, leveling off at an average of 65 per year over the following six years. These numbers, however, are deceptive. A substantial proportion of those who enrolled were only allowed to attend class-room lectures. In fact, no more than half of the students received laboratory instruction and passed the final examination. And even those who did do so suffered from a shortage of space which meant that they could work in the laboratory for only two afternoons per week.

Although the school benefited from the collection of instruments established by Ferraris, mainly for electrical measurement, the provision for experiments on electrical machinery remained modest until the end of the century. In 1900, however, the patron of the Museo and of its Scuola, the Ministry of Agriculture and Industry, offered £4,000 for the refurbishment of the laboratory. By the time the refurbishment was complete and the new premises of the Scuola e Laboratorio were opened, in 1904, the range and quality of the equipment were of a standard that would have allowed the teaching staff (if not the students) to engage in original research. Such an extension towards research had always been anticipated, for one of the distinctive features of the Scuola e Laboratorio di Elettrotecnica, undoubtedly influenced by Ferraris's scientific reputation, was a greater emphasis on original inquiries than its Milanese counterpart ever envisaged. In practice, though, the very success of the Turinese school in attracting students meant that, despite the facilities and repeated declarations of intent by the professors, energies and resources went exclusively into instruction.

In France, too, the Montefiore model made its mark in the first French institution to be entirely devoted to the high-level training of electrical engineers. The Ecole Supérieure d'Electricité (Supélec), established as a private school in 1894, provided a specialized one-year course leading to a *diplôme d'ingénieur électricien* for students who came either from one of the prestigious *grandes écoles*, in particular the Ecole Polytechnique, or (after an entrance procedure of far greater severity) from the ranks of graduates (*licenciés*) of the faculties of science and former pupils of one of the *écoles d'arts et métiers* or other lesser technical schools.[70] Although Supélec offered workshop exercises and lectures, the most important common setting in which these disparate groups of students worked was, both symbolically and in reality, the laboratory that the Société Internationale des

70. Girolamo Ramunni and Michel Savio, *1894–1994. Cent ans d'histoire de l'Ecole Supérieure d'Electricité* (Paris, 1995), chapters 1 and 2.

Electriciens—essentially, a *French* society of electrical engineers—had opened in 1888 with the aid of the profits from the international electrical exhibition of 1881 and of donations by industrial subscribers and public bodies. It was here in the Laboratoire Central d'Electricité that the pupils of Supélec were introduced to the procedures of electrical measurement and to other techniques required in the supervision of installations. These immediately applicable skills, learned primarily in the laboratory, quickly became the stock in trade of the school, and they remained the essential hallmark of the engineers who completed their training there.

Expansion and diversification

The creation of the intensive one-year courses of the Montefiore type was important for two pressing reasons. First, the courses offered the possibility of "recycling" civil and mechanical engineers who were beginning to find it more difficult to enter the job market. Secondly, they had the less concrete but equally valuable effect of helping higher technical education to benefit from the aura of scientific prestige that was associated, in public perceptions, with the industrial applications of electricity and hence with courses in that field. The success of programs of electrical instruction, both those associated with mechanical engineering and the one-year courses, is reflected in the growth in enrollments that occurred everywhere. The expansion at the Scuola e Laboratorio di Elettrotecnica in Turin, and at the Ecole Supérieure d'Electricité (where the number of students embarking on the course increased from twelve to almost fifty a year in the first decade of the school's existence) is typical.

The steady growth in the applications of electricity both within and outside industry led, in turn, to pressure for the subject to become a normal degree-level option in the engineering schools and universities, on an equal footing with civil or mechanical engineering. Although both of the early patterns to which we have referred were tenacious survivors, there was, by the turn of the century, an unmistakable move towards dedicated three-year courses (albeit ones that almost invariably still incorporated strong elements of mechanical engineering). By 1914, both of the engineering schools of Milan and Turin had created special sections offering degrees in electrical engineering—an innovation that led the *scuole-laboratorio* to abandon the task of providing postgraduate qualifications and become the electrical laboratories of the institutions to which they were attached.

In France, the move towards full three-year courses in electrical engineering had roots in optional *cours complémentaires,* detached from the main stream of the curriculum and examinations, that began to be offered in several faculties of science from about 1890. In this, Paul Janet was a notable pioneer. From 1892, as a young professor of physics at Grenoble, he mounted evening lectures on electrical technology, drawing large audiences that soon justified the conversion of a former laboratory belonging to the department of the Isère for practical work in electricity.[71] By the time he left for an appointment at the Sorbonne in 1894, public interest in the lectures and laboratory instruction was firmly established, and it continued undimmed until the establishment of a specialized *institut électrotechnique* as a semi-independent unit in the faculty of science in 1901. Here, as in the comparable institutes that were founded in association with the faculties at Lille and Nancy in 1900 and Toulouse in 1908, the three-year course for candidates entering immediately after the *baccalauréat* had pride of place, although graduates of the Ecole Polytechnique and the other elite technical schools and holders of the *licence-ès-sciences* could be admitted directly into the final year, so giving the curricula a hybrid character.[72]

In the eyes of the professors who taught in the four *instituts électrotechniques*, the gratification of the large, and ever-increasing, numbers of students who followed their teaching was tarnished by the sheer volume of routine work that success brought with it and by the mixed quality of a significant proportion of the students, in particular the many who came (as they came to the Institut Montefiore) from Russia and the Balkans.[73] Here as in other countries, however, university administrators preened themselves on the manifest popularity of the courses and on the evidence that this provided of the relevance of the faculties of science to the well-being of the region and the nation. For them, the steady growth in the number of *diplômes d'ingénieur électricien* awarded by the institutes, which continued up to the first world war, was justification enough for what certain of the more traditionally minded professors saw as an intru-

71. Paul Janet, *Notes et souvenirs* (Paris, 1933), 125–134.

72. Recent accounts of the *instituts électrotechniques* at Grenoble, Lille, Nancy, and Toulouse by Roger Moret (on Grenoble), André Grelon (on Lille), Françoise Birck (on Nancy), and Michel Grossetti and Béatrice Milard (on Toulouse) are to be found in Badel, *La naissance de l'ingénieur-électricien* (ref. 55), 57–147. For a more general study of science in Nancy, Grenoble, and Toulouse (among a number of other cities in provincial France), see Mary Jo Nye, *Science in the provinces. Scientific communities and provincial leadership in France, 1860-1930* (Berkeley, Los Angeles, and London, 1986), esp. (on Grenoble) 78-116.

73 Fox, "Science, the university, and the state" in Gerald L. Geison, ed., *Professions and the French state, 1700-1900* (Philadelphia, 1984), 66-145, on 111–118, and "France in perspective: Education, innovation, and performance in the French electrical industry, 1800-1914," in Robert Fox and Anna Guagnini, eds., *Education, technology and industrial performance in Europe, 1850-1939* (Cambridge and Paris, 1993), 201-226, on 207–209.

sive distortion of the true functions of a faculty of science by the relentless march of utility.[74]

The popularity of the new generation of specialized three-year courses in electrical engineering was evident throughout the industrialized world between the 1890s and 1914. Everywhere, the availability of such courses came to be regarded both as a mark of academic prestige and as a clear statement of an institution's will to keep abreast of industrial developments. In the new programs there was also a broad similarity in the emphasis placed on the centrality of laboratory work and on the importance of providing premises and equipment far superior to the simple, improvised facilities that had been deemed adequate in the pioneering days of the 1880s. Everywhere, in the age of high-voltage alternating currents that was spectacularly celebrated, and essentially inaugurated, at the Frankfurt exhibition of 1891, the earlier provision came quickly to appear inadequate.

Despite the underlying common aspirations, however, substantial and escalating differences of pace and degree quickly emerged. Predictably, the U.S. and those parts of Europe where the post-Frankfurt technology and its subsequent innovations (most notably in traction and the provision of industrial power) made their greatest mark soon took a clear lead in the refitting and rebuilding of laboratories. In Europe, Germany and adjacent German-speaking countries set the pace, with laboratories of electrical engineering being brought, through the 1890s, to a standard that the pioneers of the discipline could never have anticipated. A combination of the surge in student numbers and an escalating confidence in the potential of electricity provided ample justification for the expenditure that turned the electrotechnical institutes of most Technische Hochschulen into showplaces: the sums of over £40,000 that were lavished on each of the institutes at Darmstadt and Dresden between 1892 and 1903[75] compare tellingly with the maximum of £23,000 that the Drapers' Company allocated in 1908 for the construction of a new and supposedly "state of the art" electrical laboratory at Oxford,[76] and even more strikingly with the far lesser sums ex-

74. Fox, "France in perspective"(ref. 73), 208 (Table 8.1).
75. See Lexis, *Die Technischen Hochschulen* (ref. 18), 242 and 284. The expenditure at Darmstadt allowed the 25th anniversary of the founding of the Elektrotechnische Institut there, in 1907, to be celebrated with ostentatious but justified satisfaction; see *Das Elektrotechnische Institut der Grossherzoglichen Technischen Hochschule Darmstadt. Zur Feier seines 25 jährigen Bestehens* (Darmstadt, 1907).
76. The Drapers' Company made available a maximum of £22,000 for the construction of the new laboratory and £1000 for "furniture and equipment." See *Oxford University gazette*, *39* (10 Oct 1908), 89. The completed building was handed over to the university in 1910.

pended during the 1890s on the new electrical engineering laboratories at University College and King's College, London.[77]

As a result of this escalation of expenditure in the 1890s and the early years of the twentieth century, the gulf between the material facilities for the study of electricity in the engineering schools of Germany, Switzerland, and (more unevenly) the Austro-Hungarian Empire and those in the rest of Europe first became marked. In the countries of the German-speaking area that found themselves in the "fast lane" of technological and economic development, not only opulent buildings but also new chairs abounded.[78] An important stimulus for this expansion came from the Chicago World's Columbian Exposition in 1893, where the spectacular advances of the American electrical industry provoked anxious German reflections on the material provision in the Technische Hochschulen, good (or at least adequate) though this was by the standards of other parts of Europe.[79] Action followed quickly. At the Hochschule in Berlin, whose existing buildings, opened in 1884 at Charlottenburg, already inspired envy within and beyond the Empire, a new wave of spending on laboratories and other facilities for electrical and mechanical engineering, including a purpose-built electrochemical laboratory and an associated chair in the subject, helped to justify the unbridled outpouring of pride at the school's centenary celebrations in 1899.[80] At about the same time at Darmstadt, Kittler's teaching was reinforced by a second chair of *Elektrotechnik*, created in 1894,

77. The laboratory at University College was set up with the aid of £6000 that Lady Siemens gave to the college in 1890. Of the £7300 made available to King's College as a memorial to the engineer John Pender, only £5000 was allocated for the laboratory, which remained meagerly supported.
78. Fox and Guagnini, "Life in the slow lane: Research and electrical engineering in Britain, France, and Italy, ca. 1900," in Peter Kroes and Martin Bakker, eds. *Technological development and science in the industrial age. New studies on the science-technology relationship* [Boston studies in the philosophy of science, volume 144] (Dordrecht, Boston, and London, 1992), 133-153, on 134–137.
79. Karl-Heinz Manegold, *Universität, Technische Hochschule und Industrie. Ein Beitrag zur Emanzipation der Technik im 19. Jahrhundert unter besonderer Berücksichtigung der Bestrebungen Felix Kleins* (Berlin, 1970), 116–120 and 147.
80. The splendor of the celebrations, which took place over four days and were attended by the Kaiser, is brought out well in two sumptuous volumes recording the events: Alfred G. Meyer, *Die Hundertjahrfeier der Koeniglichen Technischen Hochschule zu Berlin. 18.–21. October 1899* (Berlin, 1900), and *1799–1899. Festschrift der Hundertjahrfeier der Königl. Techn. Hochschule zu Berlin* (Berlin, 1903). For a self-congratulatory account of the expansion and investment in buildings and facilities between 1884 and 1899, see also *Chronik der königlichen Technischen Hochschule zu Berlin, 1799–1899* (Berlin, 1899), 123–269.

as well as by the expenditure on buildings and fittings to which we have already referred.[81] Similar patterns of investment and new appointments of either full professors or *Privatdozenten*, as well as rather large numbers of assistants, brought the institutes or electrical sections at Hanover, Karlsruhe, Dresden, Aachen, Munich, Braunschweig, and Stuttgart to a greatly improved standard.[82] And these developments were amply matched beyond the boundaries of the Empire. The ETH in Zurich created three new professorships, with associated facilities, in branches of electrical technology between 1895 and 1905.[83] And at the Technische Hochschule in Vienna an increase in the number of chairs in the subject from one to four between 1903 and 1908, allied to the inauguration of a much-needed new building (costing £37,500) in 1904, suddenly propelled the school to the forefront of technical institutions in central Europe.[84]

Despite the unprecedented intensity of construction and refurbishment and the general association of electrical technology with the idea of progress and rapid change, the emphasis in departments and laboratories remained firmly on teaching and only to a much lesser extent on the contribution that academic laboratories might make to innovation. The fact that the industrial applications of electricity were perceived as closely related to work at the frontiers of science might have been expected to give research a prominent role. But this was no more the case than it was in mechanical

81. König, "Kittler" (ref. 44), xxxiv.
82. Lexis, *Die Technischen Hochschulen* (ref. 18), 137–146 and 181–301, and König, *Technikwissenschaften* (ref. 40), 13–86, are excellent general sources on these and related developments. For a contemporary foreign view of the growth in chairs and facilities in the Technische Hochschulen, see Albert Granger, *Etude de quelques laboratoires industriels des écoles techniques supérieures en Allemagne* (Paris, 1901). More detailed accounts of the various new installations and appointments in electrical technology in the 1890s and early years of the new century are also plentiful. See, for example, in addition to the items cited in refs. 34 and 79, *100 Jahre Technische Hochschule Hannover. Festschrift zur Hundertjahrfeier am 15. Juni 1931* (Hanover, 1931), contributions by various authors on pp. 232–240; and Frank Stark, ed., *Die k.k. Deutsche Technische Hochschule Prag 1806–1906. Festschrift zur Hundertjahrfeier* (Prague, 1906), 407–411.
83. Lasius, "Die Gebäude der Eidgenössischen Polytechnischen Schule" (ref. 47), 2, 336–343, and Dünner et al., "Die Abteilung für Elektrotechnik" (ref. 46), 424–430.
84. Wolfgang König, "Vom Schlußlicht zum Spitzenplatz. Die Elektrotechnik an der TH Wien 1882–1914," *Blätter für Technikgeschichte, 55* (1993), 41–63, on 46–53. See also the contributions to mark the fiftieth anniversary of the opening of the Elektrotechnisches Institut, in *Elektrotechnik und Maschinenbau, 71*. Jahrgang, Heft 9 (1 May 1954), and Richard H. Kastner, "Die Technische Hochschule in Wien. Ihre Gründung, Entwicklung und ihr bauliches Werden," in H. Sequenz, ed., *150 Jahre Technische Hochschule in Wien 1815–1965* (2 vols., Vienna and New York, 1965), 2, 77–80.

engineering, and in higher technical schools research remained a secondary activity. Professors might pursue it in a personal capacity, usually (in the manner of John Ambrose Fleming, Gisbert Kapp, Jules Jamin, and Galileo Ferraris) as consultants to electrical companies; and a few selected students might even be cautiously engaged as assistants. These individual instances, however, though not rare, do not undermine the point that in electricity, as in other areas of engineering, research was institutionally a late-comer whose place in higher technical education was not recognized until after 1900.

Academic research in electrical engineering

By the beginning of the twentieth century, in electrical engineering as in the other scientific and technical disciplines, it was a matter of pride for academic institutions to provide, and parade, the most opulent facilities they could afford and to offer the possibility of an engagement in research as a means of attracting students. Yet no degree of opulence could change the view, held by manufacturers as well as by teachers of engineering, that, despite a few conspicuous exceptions (such as the tantalum filament lamp), the quest for originality and the pursuit of novel lines of inquiry were best left to engineers working outside the academic world. The function of laboratories, by contrast, was seen to lie mainly in the provision of training in the indispensable but far less creative tasks of testing, calibration, and the assessment of machinery (albeit in trial runs on a scale very different from that of a true industrial plant and in an environment that could not convey the complexity of the contexts in which electricity was commercially generated or used).

The national research laboratories that were set up, with state support, from the late nineteenth century enshrined a similar orientation towards routine services rather than innovation. In this, the best-known of them, the Physikalisch-Technische Reichsanstalt, inaugurated in 1887 in Berlin, was exemplary, being in important respects a typical product of what Peter Lundgreen has identified as the rise of the "regulatory state."[85]

85. Peter Lundgreen, *Standardization—Testing—Regulation. Studies in the history of the science-based regulatory state (Germany and the U.S.A., 19th and 20th centuries)* [Wissenschaftsforschung/Science studies, Report 29] (Bielefeld, 1986), containing two studies: "Normierung und Kontrolle: der Bereich 'Technik' in der staatlichen hochschulfreien Forschung (Deutschland, 1870–1980)" and "Measures for objectivity in the public interest: The role of scientific expertise in the politics of technical regulation: Germany and the U.S., 1865–1916." For a French view of the value to industry of the German Empire's provision for testing laboratories on the eve of the first world war, see J. Roux, *Les laboratoires industriels d'essais en Allemagne* (Paris, 1913).

Lundgreen's characterization stands despite the fact that, as its founder Werner von Siemens originally conceived the PTR, the balance between research in experimental physics and in technology was to be tipped unmistakably in favor of the former.[86] Siemens, who had studied at the Artillerie- und Ingenieurschule in Berlin in the 1830s, had always cherished the scientific component of his training and maintained a strong commitment to pure science. Hence his vision of the PTR's potential as a contributor to the imperial economy went hand in hand with a conviction that his own engagement with science (both through the research he performed and through his contacts with scientists) had been the bedrock of his industrial success. Initially, Siemens was able to impose his philosophy: the allocation of £35,000 in the Reichstag (including £24,000 for the construction and fitting out of the scientific section) reflected the success his campaign enjoyed, in the mid-1880s, in the face of political opponents who wished to see an institution pursuing only technical work of immediate industrial relevance.[87] The appointment of the eminent but now aging Hermann von Helmholtz as the PTR's first president was another gauge of Siemens's victory, one that laid the foundations of the institution's reputation as a leader in physical metrology

The supporters of a more marked industrial orientation did not lapse into inactivity, however, and the primacy of science was soon eroded. In the 1890s, the pace of work in the laboratories of the Technical Section outstripped that in the Scientific Section, and, more crucially, the involvement of both sections in routine testing for industrial clients, in particular in the Technical Section's electrical laboratory, assumed a position so dominant that it stretched resources to a dangerous degree.[88] Moreover, even when fundamental scientific research of great importance was performed, as in the case of the study of black body radiation by Otto Lummer, Ernst Pringsheim, and others between the mid-1890s and the early years of this century, the main justification for the work was at least as utilitarian as it was scientific.[89]

86. David Cahan, *An institute for an empire. The Physikalisch-Technische Reichsanstalt 1871–1918* (Cambridge, 1989), 35–58.
87. Ibid., 51. The contribution of £24,000 represented just under a fifth of the total expenditure on the construction and fitting out of the PTR between 1887 and 1897. Cahan (ibid., 103) gives figures of £48,000 and £135,000 for the sums expended on the establishment of, respectively, the scientific section and the technical section in this period (in addition to the running costs and land, donated by Siemens, with a value of over £28,000).
88. Ibid., 88–121 and 206–223.
89. Ibid., 143–157.

In Britain, the profile of the activity of the National Physical Laboratory, created at Teddington, on the Thames to the west of London, in 1900, was even more resolutely that of a service institution.[90] As early as 1868, at a meeting of the British Association for the Advancement of Science, Lt. Col. Alexander Strange, a Fellow of the Royal Society and former employee of the Indian Trigonometrical Service, argued for "the necessity for state intervention to secure the progress of physical science," in the form of new national institutions for research, as a way of securing knowledge that would eventually help industry to respond to the growing competition between European countries.[91] The nature of the research that industrial companies regarded as most appropriate for such state-funded agencies was articulated clearly by the electrical manufacturer Col. R.E. Crompton. For Crompton, speaking before the Society of Telegraph Engineers and Electricians in 1885, the need for "a national establishment for standardising electrical measuring instruments" was so great that if the state did not act, industrialists would have to pool their resources to do so.[92] Physicists, for their part, were even keener than manufacturers to have a new institutional setting for research. In pursuit of their objectives, they eagerly deployed arguments based on the practical value of such an institution, and, in fact, it was from within the scientific community that the cause of a government-funded physical laboratory drew renewed strength in the crucial decade beginning in 1890. Again, the immediate context was a meeting of the British Association for the Advancement of Science, held on this occasion in Cardiff in 1891 and addressed, in his capacity as president of the section for mathematical and physical science, by the professor of physics at University College, Liverpool, Oliver Lodge. As Lodge argued, without a national laboratory in which a tradition of accuracy and reliability could be established, Britain could never hope to take her "proper share of the

90. Edward Pyatt, *The National Physical Laboratory. A history* (Bristol, 1983), 12–61 (on the period to 1914), and Russell Moseley, "The origins and early years of the National Physical Laboratory: A chapter in the pre-history of British science policy," *Minerva, 16* (1978), 222–250.
91. The words quoted are the title of a paper that Strange, a prominent early spokesman for the cause of science, read at the 1868 meeting of the BAAS at Norwich. See *Report of the thirty-eighth meeting of the British Association for the Advancement of Science; held at Norwich in August 1868* (London, 1869), "Notices and abstracts of miscellaneous communications to the sections," 6–8.
92. *Journal of the Society of Telegraph-Engineers and Electricians, 14* (1885), 510–511. Crompton was speaking in the discussion following J.A. Fleming's paper "On the necessity for a national standardising laboratory for electrical instruments," ibid., 488–503.
93. *Report of the sixty-first meeting of the British Association for the Advancement of Science held at Cardiff in August 1891* (London, 1892), 547–557, esp. 550–551. Lodge

highest scientific work of the world," with all the attendant benefits for the "arts" that were presented as axiomatic.[93]

Through the 1890s, following Lodge's address, a number of committees were organized with the task of promoting and defining the scope of the proposed laboratory. Backed by the Royal Society, with the support of professors of physics from the leading institutions of higher education, and by the administrative and tactical acumen of the retired military engineer and former secretary of the British Association for the Advancement of Science Sir Douglas Galton, the plan finally won governmental approval in 1899.[94] Immediately, formal responsibility for the control of the laboratory was vested in a governing body, on which the Royal Society had a preponderant voice,[95] and Richard Glazebrook, who had been a lecturer in mathematics and Fellow of Trinity College in Cambridge, was appointed its director. There could scarcely have been a more appropriate choice for the directorship. Glazebrook came with all the right credentials, including experience as secretary of the British Association for the Advancement of Science's electrical standards committee, and with the influential support of Lord Rayleigh, the former Cavendish professor of experimental physics and director of the Cavendish Laboratory in Cambridge (where Glazebrook had been assistant director for some years in the 1890s). He was to serve for twenty years.

From the start, the utilitarian thrust of the National Physical Laboratory's work was determined as much by necessity as by choice. The promoters of the institution had always believed that the NPL should reserve at least part of its activity for more fundamental types of applied research. But the conditions imposed by the Treasury left little scope for work of this kind. When the project was launched, the financial support offered by the government consisted of an initial capital expenditure of £17,000 and an annual grant of £4,000 for five years; after that, the laboratory was expected to become self-supporting, the necessary income being raised from fees for testing. In the event, the annual grant was renewed and, from 1909, raised to £7,000, but the laboratory continued to depend heavily on the fees charged

---

was a committed advocate of Maxwell's electromagnetic field theory who had also performed important work on the industrial applications of electricity. He held a number of patents, both individually and jointly with his partner, the scientific instrument-maker Alexander Muirhead.

94. At the time of his involvement in the planning of the National Physical Laboratory, Galton, who was a Fellow of the Royal Society, was also chairman of the management committee of the Kew Observatory.

95 On the Royal Society's role in the moves that led to the creation of the NPL, see Moseley, "The origins and early years of the National Physical Laboratory" (ref. 90), 231–233.

amounted to £13,500 in 1909 and £30,245 in 1914, most of it earned by the metrology and electrotechnics departments, for the ever-growing volume of work on electrical standards, the testing of electrical instruments, and the measurement and recording of temperature, and by the engineering department, for studies of fatigue phenomena, in particular the effects of wind on bridges and roofs.[96] Private donations, too, were an essential means of survival. Some of these, usually earmarked for special projects, were substantial: Alfred Yarrow's contribution of £20,000, which helped the laboratory to construct a 550-foot-long tank for experiments with scaled-down ships, is a good example.[97] Nevertheless, the laboratory remained the poor relation of the Physikalisch-Technische Reichsanstalt, suffering both from the relatively modest expenditure on its initial establishment and from the size of its regular governmental grant, which was smaller than that which the German government made to the PTR.[98]

In France, the position of laboratories active in testing and calibration was as precarious as it was in Britain. The Laboratoire Central d'Electricité in Paris, for example, could not have survived without constantly selling its testing and metrological services, mainly to private firms.[99] In this respect, it resembled the testing laboratories in civil and mechanical engineering, such as the materials testing laboratory of the Ecole des Ponts et Chaussées and the Laboratoire d'essais mécaniques, physiques et chimiques et de machines at the Conservatoire National des Arts et Métiers, both of which were run essentially as commercial enterprises, despite their almost official national status.[100]

96. Pyatt, *The National Physical Laboratory* (ref. 90), 40–51. On the income and expenditure of the NPL, see also Moseley, "The origins and early years of the National Physical Laboratory" (ref. 90), 239.
97. Yarrow owned one of the most successful shipyards in Britain. His pioneering work on the application of advanced technology to the design and construction of ships, esp. for military use, led to his election as a Fellow of the Royal Society in 1922.
98. Cahan, *An institute for an empire* (ref. 86), 133. Cahan notes a six-fold disparity between expenditure on the PTR and the NPL in their early years. As Moseley has observed, however, the British government's reluctance to provide funding for industrial research was tempered by its interest in work with a military purpose. The sums allocated for aeronautical research, for example, rose from £1,524 in 1909 to £5,311 on the eve of the first world war. See Moseley, "The origins and early years of the National Physical Laboratory" (ref. 90), 239 and 244–245.
99. Ramunni and Savio, *Cent ans d'histoire de l'Ecole Supérieure d'Electricité* (ref. 70), 31–34. On the precarious finances of the laboratory at its foundation in 1888 and in its early years, see André Grelon, "La formation des ingénieurs-électriciens," in François Caron and Fabienne Cardot, eds., *Histoire générale de l'électricité en France. Vol. 1. Espoirs et conquêtes 1881–1918* (Paris, 1991), 254–291, on 286–293.
100. For a vivid impression of the effects of the precariousness of the laboratory at

In the French provinces, the pressures to seek commissioned work were as great as they were in Paris, and the opportunities were at least as abundant, arguably even richer. Here, the laboratories of the four provincial *instituts électrotechniques*, at Grenoble, Nancy, Lille, and Toulouse, though not enjoying the formal national standing of the Laboratoire Central d'Electricité, all substantially augmented their income by providing services, mainly to their local communities. At Grenoble, in particular, the Bureau de contrôle et d'essais industriels (later the Bureau d'essais électriques et d'étalonnages industriels) became, from its inauguration in makeshift premises in 1900, the core of a considerable business enterprise.[101] Acting both as a source of laboratory services for the town's system of electrical supply and as an increasingly valued resource for private electrical manufacturers and municipal installations, it carried out a volume of commissions (chiefly to do with the calibration of meters, tests on motors and generating equipment, and advisory work on safety and efficiency) that soon made the quest for significant expansion a recurring necessity. By 1913, it engaged not only its own chief engineer but also six other senior technical staff, in addition to mechanics, assistants, and part-time advisers, and did so in an elaborate, though barely adequate set of rooms devoted to the housing of accumulators, trial runs of machinery, calibration, precise measurement (in particular of resistance and capacity), and photometry, as well as to workshops and offices. In the first twelve years of its existence, it performed over 25,000 tests on meters alone, providing certificates authorized by the Ministry of Public Works.[102] The opinion of the director of the institute, Louis Barbillion, that this work constituted "the main factor in the attraction we have for industrialists" was no exaggeration.[103] Without

---

the CNAM, see Henry Couriot, "Conservatoire National des Arts et Métiers (1913–1919). Rapport général du Conseil d'Administration du Conservatoire National des Arts et Métiers sur l'état de cet établissement, le fonctionnement de ses services et les résultats de l'enseignement," *Journal officiel de la République Française*, Annexe (feuille 2), 13 Jan 1921, 17–28, esp. 19, 22, and 24–6. On the work performed at the Ecole des Ponts et Chaussées, which embraced both physical tests on metals and other materials and chemical tests, notably on cement, see *Ministère des Travaux Publics. Laboratoires de l'Ecole Nationale des Ponts et Chaussées. Note sur leurs origines, leurs installations, les appareils et méthodes d'essai employés et leurs travaux* (Paris, 1891), an account that stresses the poor standard of the facilities.

101. For a good contemporary account, see Louis Barbillion, *Le bureau de contrôle et d'essais industriels annexé à l'Institut polytechnique de l'Université de Grenoble* (Grenoble, 1913).

102. Ibid., 14.

103. The phrase is in Barbillion's "Note sur le maintien du Bureau de contrôle et d'essais sur les emplacements actuels," a typescript addressed to the Rector of the

its commissions, the institution would have struggled as a modest vehicle for teaching, with, at best, a local reputation and facilities far short of those it came to enjoy after the major rehousing that brought together the Institut Electrotechnique and the Ecole Française de Papeterie (with its own very important testing laboratory) in a unified Institut Polytechnique shortly before the first world war.[104]

It was by no means uncommon for work of the kind that was conducted at Grenoble to be referred to as research. But this reflected the range of activities that could be embraced by that term and the divergence of opinion on the meaning and value of research in higher education that had emerged from the wider debate on the role of laboratories since the 1880s. Arguing on pedagogical grounds similar to those advanced some years earlier in support of the comparable case in the physical sciences, the champions of research (who were vociferous though never numerous) suggested that engineering students would benefit from an involvement in some form of original work in later stages of their education; this involvement, as the champions of research insisted, was important not only those who aspired to academic careers, but also for those preparing for positions of responsibility in industry. However, the question of what form the original work should take proved even more controversial than it had been in the case of "pure" science. The issue was addressed in 1884, in a characteristically straightforward way, by Fleeming Jenkin, professor of engineering at the University of Edinburgh from 1868 to 1885. In pursuit of his ideal of an extensive common ground where the interests and skills of the teacher of science would come into contact with those of the practical man, Jenkin urged that the initiation in laboratory work should take the form of "measurement classes."[105] In his view, it was precisely the day-to-day habit of precision that should be impressed on the mass of students who were to become the rank-and-file engineers so much needed in modern industry. While he conceded that a few students would emerge with "a great natural aptitude for scientific inquiry,"[106] he argued that the majority would, and should, leave their student laboratories with skills and experience of a very different kind. Citing specifically the needs of the "second-best young man," he stated:[107]

---

academy of Grenoble and composed (apart from some later annotations) in Sep 1909, in Archives Nationales, F[17] 14502.

104. *Institut Polytechnique de l'Université de Grenoble, 1900-1950* (Grenoble, 1950), 72–103.

105. Fleeming Jenkin, "On science teaching in laboratories," in Jenkin, *Papers literary, scientific &c.*, Sidney Colvin and J.A. Ewing, eds. (2 vols., London, 1887), 2, 183–190.

106. Ibid., 184.

107. Ibid., 185.

> We cannot give him the hunger for knowledge, the acute logical discrimination, nor the imaginative faculty required for research; but we can teach him how to ascertain and record facts accurately; we can bring home to him the truth that no scientific knowledge is definite except that based on the numerical comparison which we call measurement; we can teach him the best modes of making that comparison in respect of a vast number of magnitudes, and in teaching this we shall teach him to use his hands and eyes. This practical teaching gives clear conceptions to the minds of many who receive a verbal definition as a mere string of dead words.

Over the next fifteen years, this approach, with its roots in precision measurements and a close association with the low-current technology of telegraphy, was powerfully influenced by the new industrial applications of electricity. As a result, the emphasis shifted. Increasingly, engineers involved in the design and adaptation of electrical machinery and plants found themselves having to tackle problems of a far from routine character, ones that called on them to adapt existing technologies to novel, unexplored contexts, mainly in manufacturing industry. As we point out in the next chapter, such work required not only a solid educational background and experience but also a significant degree of originality, albeit of a directly applicable kind. The research that began to appear in the programs of engineering schools towards the end of the century reflected this reality. It still did not imply the pursuit of fundamentally novel lines of inquiry, but it did imply an engagement with practical problem-solving, often of a very demanding nature.

The changing realities of the work of a practicing electrical engineer in the 1880s and 1890s made their mark on the institutions of higher technical education, at the same time as research and laboratory facilities were coming to be perceived as a source of status across the whole range of the engineering and other applied disciplines. As early as the 1880s, students at the Institut Montefiore were required, in the later stages of their course, to engage in a carefully focussed independent project; in this way, it was said, they would be able to "exercise their sagacity and hence develop their spirit of initiative and invention," albeit in a well-bounded context.[108] And slowly and cautiously professors everywhere became readier to involve able pupils not only in their own commissioned work but also in investigations that began, even more slowly, to have a more open-ended character.

108. *Institut Electro-technique Montefiore annexé à l'Université de Liège* (ref. 57), 5.

A significant impediment to this trend, especially in countries such as Britain, where college fees were high, was the difficulty of inducing students to stay on for one or more years after taking their degree in order to engage in research. As in the case of the science departments, however, the developing system of demonstratorships (introduced as a way of finding teachers to cope with rising numbers, in particular in laboratory classes) provided some financial underpinning and incentive. The case of William E. Sumpner, who studied in the Department of Physics and Electrical Engineering at the Central Institution and received the diploma of Associate of the City and Guilds of London Institution (essentially a bachelor's degree) in 1887, shows how effectively the system could work. Working from 1887 to 1893 as a demonstrator under Ayrton, Sumpner had time to carry out what was described as "much original and valuable research work," to read papers to eminent societies, and to publish.[109] His experiments, as would be expected of someone close to Ayrton, were firmly in the field of electrical engineering, but when in the late 1880s he wished to proceed to a higher degree, he was unable to do so in his chosen discipline, in which the doctorate did not yet exist. He therefore had to follow the only track open to him and in 1889 he took his London University D.Sc. in experimental physics.

Through the 1890s, at least in Britain and Germany, the signs of a growing institutional openness to the principle of formal mechanisms and rewards for research performed by advanced students of engineering multiplied. In Britain, the first grants for students pursuing technical research became available in the mid-1890s. At the City and Guilds Central Technical College, for example, funds from the Salters' Company (one of several City of London livery companies that supported the City and Guilds of London Institute) went, in 1894, to the creation of a fellowship of £150 p.a. for advanced research in some area of industrially related chemistry, and two years later a second fellowship in the subject was financed by the Leathersellers' Company, encouraged by its Master, the chemist W.H. Perkin, Jr.[110] Despite their modest scale, the initiatives were warmly welcomed within the college, which saw student research and publication as a source of both institutional distinction and professional advantage. The Report of the Council for the year 1890 was explicit in its suggestion that

109. *City and guilds of London Institute. Report of the Council upon the work of the Institute for the year 1894* (London, 1895), 4–5.

110. Ibid., 1–2, and the corresponding report on 1896 (London, 1897), 3. In the context of this development, we use the form City and Guilds Central Technical College, which had replaced the name City and Guilds Central Institution in 1893; see above, 114.

the success of the college's students in filling important positions in industry owed much to their engagement in independent advanced work:[111]

> One of the causes that have contributed to these early successes is due, no doubt, to the encouragement given, especially to third year students, to undertake or assist in some line of original research. The importance of this branch of the work cannot well be over-estimated. New processes and new methods are constantly applied to all our manufactures effecting better or more economic production, and it is essential that those who may in future years take a prominent position in directing the many branches of industry carried on in Great Britain, shall have that spirit of research and that inventive genius which will enable them to take hold of and apply new discoveries in pure science.

Six years later, the report of a special committee, appointed by the governors, referred to what had happened, across the whole range of the college's disciplines, with pride and probably some exaggeration:[112]

> In considering the educational work of the College, we must not omit to mention the large number of contributions to the advancement of science which have been made by the Professors individually and in co-operation with their students. Indeed the spirit of research pervades every department of the College.

In electricity, the investigations conducted or completed in 1896–1897 included "Oscillographs," "On a determination of the Ohm made in testing the Lorentz apparatus of the McGill University, Montreal," "Variations in the E.M.F. of the H. form of Clark cells with temperature," and "Some tests on the variation of the constants of electricity supply meters with temperature and with current," all of which, if not particularly adventurous, entailed the application of known methods to new problems.[113]

The Manchester Technical School was typical of the many other schools in which a growing concern with research marked the change of status from that of a purely vocational institution. The process of upgrading, which began in 1890 and culminated in the school's designation as the Faculty of Technology of the University of Manchester in 1904, was also marked by a gradual loss of interest in part-time education and a shift of emphasis from

111. *City and guilds of London Institute. Report of the Council upon the work of the Institute for the year 1890* (London, 1891), 4.
112. *City and guilds of London Institute. Report of the Council upon the work of the Institute for the year 1896* (London, 1897), 25.
113. *City and guilds of London Institute. Report to Governors. Report on the City and Guilds Central Technical College, 1896–97* (London, 1897), 8.

evening classes to full-time day courses. By 1900, the school had well-equipped laboratories, and plans were in place for a new building with even better facilities. Amid these changes and conflicting pressures, the commitment to instruction that was oriented to the needs of industry remained a distinctive feature. But even here the academic staff were encouraged to pursue research, so long as the goal was the solution of the practical problems of industry. Students too had the opportunity of embarking on more advanced work, and from 1908 the Master's degree by research, M.Sc. (Tech), was offered for a one-year thesis that presented "a record...of original work, or...an essay on some scientific question of importance involving criticism or historical research."[114] To support the students working for the degree, the City of Manchester's Education Committee began, in 1910, to offer five studentships for advanced study and research each year.[115] Encouraged by this show of confidence and the school's sustained commitment to the initiative, 65 students were awarded the degree (17 of them in electrical engineering, compared with 31 in applied chemistry) between 1908 and 1913. To what extent the municipal grants were actually used by the recipients in order to pursue what could properly be described as independent lines of inquiry (as opposed to following a course of instruction) is unclear. What is beyond question, however, is that in the annual calendars and elsewhere much was made both of the M.Sc. (Tech) and of the research performed by the staff, though always with the gloss that students working for the degree did so after a period spent in industry. British Westinghouse and Mather and Platt were two of the prominent local firms where candidates often pursued a postgraduate apprenticeship before taking the M.Sc. (Tech) and then, in many cases, returning to work once the degree was completed.

In the more research-oriented culture of Germany, professors in the Technische Hochschulen had traditionally shown an even greater sensitivity than their peers in Britain to the importance of research, especially in pursuit of their quest for professional status. The challenge they now conceived—more pressingly than ever—was that of harnessing this research culture for the academic advancement of their institutions. The main focus of their strategy was the qualification of the doctorate, which since the 1880s

114. *Report upon the necessity of a new and efficient organisation of the new Municipal Technical School*, May 1901, 13. The document, a printed report of the sub-committee appointed to review the structure of the school on its installation in the new building in Sackville Street, is in the Archives of the University of Manchester Institute of Science and Technology (BR 607.40 MAN). It is marked "Strictly private and confidential."
115. *Manchester Municipal School of Technology. Calendar 1914–5* (Manchester, 1914), 177–178.

they had wished to see bestowed on their most advanced students on the same terms as those that applied to students in the universities.[116] However, when the struggle for the doctorate, along with the associated right to grant the title of "Diplom-Ingenieur" to all of their graduates, was eventually won, following the decisive intervention of the Kaiser in 1899, the victory was slightly soured by the adoption of the term "Doktor-Ingenieur" to distinguish the new degree from the doctorate offered by the universities. Even the designated abbreviation "Dr.-Ing." had to be printed in gothic as opposed to the roman script used for the university degree. More seriously, although the doctorate in engineering (*Doktorat der technischen Wissenschaften*) was offered by Technische Hochschulen throughout the German and Austro-Hungarian Empires by 1901, and at the ETH in Zurich from 1909, it was slow to catch on.[117] At Aachen, only 37 doctorates were awarded between 1904 and 1921 in the section for *Maschineningenieurwesen*; of these no more than two or three had any bearing on electricity. The situation at the German Technische Hochschule in Prague, where four of the 34 doctorates awarded in the school between 1901–2 and 1905–6 seem to have been for electrical research, was somewhat better. But even here research in the sense of original work could not be said to have become a central element in the *cursus* of students of electrical engineering.[118]

One reason for the muted response to the new doctorate was the resurgence of the economic recession between 1901 and 1903, which had the effect of suddenly diminishing confidence in career prospects, especially in electrical engineering, and hence, in most Technische Hochschulen, of slowing or even reversing the flow of students specializing in *Elektrotechnik*.

116. Manegold, *Universität, Technische Hochschule und Industrie* (ref. 79), 249–305.
117. A vivid impression of the work done for the new doctorate in its early years is given in Paul Trommsdorf, *Verzeichnis der bis Ende 1912 an den Technischen Hochschulen des Deutschen Reiches erschienen Schriften* (Berlin, 1914), where all the publications emanating from the Technische Hochschulen, including the titles of the doctoral dissertations, are listed. A statistical summary of the number of doctoral dissertations written and the various Technische Hochschulen is given in König, *Technikwissenschaften* (ref. 41), 262–263. König's figures and discussion bring out the difficulty of determining precisely how many doctorates were awarded. But the point about the modest numbers, of roughly ten a year in *Elektrotechnik* for the whole system of eleven German Technische Hochschulen, stands.
118. At the Technische Hochschule of Braunschweig, 112 doctorates were awarded between 1901 and 1914; the great majority were in the chemical section, and only 29 were given for the whole Machinenbauabteilung, embracing not only electrical engineering but also mechanical engineering and textiles; see Helmuth Albrecht, *Technische Bildung zwischen Wissenschaft und Praxis. Die Technische Hochschule Braunschweig 1862–1914* (Hildesheim, 1987), 349.

But a profound source of reticence also lay in the lingering ambivalence concerning the place that research in both electrical and mechanical engineering should have in higher technical education. With the campaign for academic respectability won, and the doctorate in engineering a reality, divisions that had simmered in the 1890s re-emerged. Now opposition to the new degree came from two very different quarters. On the one hand, there was the view, given substance by the opening of Felix Klein's Institut für technische Physik at the University of Göttingen in 1896, that original research—even in such an industrially relevant area as applied thermodynamics, which became a Göttingen speciality—was properly the exclusive preserve of the university.[119] On the other hand, there were many in the industrial community who remained convinced that whatever value laboratory exercises might have in the inculcation of applicable professional skills and habits of precision, important innovations in electrical technology were less likely to emerge from research in a university or a technical school, however well equipped, than from the sites where high-voltage alternating current was making its mark on the realities of everyday life.

There was much substance in this latter view. By the early years of this century, in fact, the most interesting "laboratories" contributing to the diversification of the applications of electricity were in the factories, generating stations, and supply and communication networks, where innovation resulted from an engagement with working machinery and installations. It was primarily in this world of full-scale test-runs, in conditions far removed from the idealized circumstances of the laboratory, that the real excitement of electrical technology lay in the quarter of a century before the first world war.

119. Manegold, *Universität, Technische Hochschule und Industrie* (ref. 79), 162–168.

It is not difficult to understand why research carried out in formally designated laboratories has assumed such a central position in historical studies of science-based industry. Viewed through the distorting prism of the period since 1918, in which an ever-increasing number of large firms have devoted substantial resources to research, the laboratory has come to be conceived as a principal source of technological innovation. The danger of such a perception is obvious. When projected on to the period before the first world war, it tends to impose a selective and limiting vision both of the importance of laboratory research and of what constituted research. The point is not that this perception is wholly false; it was in this earlier period, in fact, that the first in-house research laboratories in the modern sense were created—mainly in Germany in the chemical industry and in Germany and the U.S. in the electrical industry. The problem is rather that because of their pioneering character, this first generation of laboratories has occupied a favored position in the attentions of historians and that, as a result, studies of innovation have tended to concentrate on them to the exclusion of other motors of technological change with which this chapter is chiefly concerned.

Here, we develop two lines of interpretation based on the broader perspective we have just advocated. One line of interpretation arises mainly from recent studies of Germany and the U.S., which suggest that in reality even the work carried out in the first designated "research laboratories" preserved strong elements of testing, quality control, and investigations directed at the protection of patents, rather than a preoccupation with the development of new products and processes.[1] There can be no denying that these activities were economically profitable, and they were ones for which graduates who had undergone academic training in laboratory techniques were well suited. The fact remains, however, that they were far removed from the commitment to the pursuit of novel lines of inquiry that most universities and higher technical schools were trying to foster as at

1. See the studies by Meyer-Thurow, Homburg, Reich, Reinhardt, and Wise, cited below in refs. 12, 13, 18, and 54.

least one facet of their attempts to promote their commitment to science and more generally the modernization of learned culture.

Our second line of interpretation, which is linked to the first, also stresses the need to cast our net widely in search of the sources of technological change in the later nineteenth century. Independent inventors continued to have their role, as we readily acknowledge, but even when that role is recognized, the origins of the remarkable train of innovations that characterized the electrical industry from the 1880s remain to a large extent unidentified. Faced with this challenge, we use this chapter to bring out the undervalued importance of experimental procedures conducted not in laboratories specially set up for the purpose but rather on the shop floor and the site by technical staff engaged in the day-to-day work of design, production, and installation. This emphasis entails a more capacious conception than is usual of the nature of research in the period, but it is a conception to which the history of the electrical industry (as also of the metallurgical and machine construction industries) leads inexorably.

Unraveling the rhetoric

The statements that did most to give laboratory-based research the prominence it had, and still has, in explanations of industrial success before the first world war were stimulated initially by the rise of Germany as a scientific and industrial power. Not surprisingly, it was in France, humiliated in 1870 and ever-watchful of developments over the Rhine, that the preoccupation with German might emerged most strongly as a standard literary trope. There, from the 1870s, an earlier, predominantly academic admiration for the quality of German laboratories and the critical spirit of German intellectual life that had been commonly voiced in liberal circles hostile to the Second Empire quickly found its way into a far wider current of envy, soul-searching, and *revanchiste* sentiment that often had industry as a prominent focus.[2] Almost inevitably, it was with respect to the newly burgeoning sector of artificial dyestuffs and other fine chemicals that the point first entered public debate. And it is for this reason that we return initially to chemistry and the chemical industry, before discussing electricity and some other areas of production to which attention spread during the last quarter of the nineteenth century—a period in which the U.S. matched and, in cer-

2. In an abundant secondary literature, see, for example, Harry W. Paul, *The sorcerer's apprentice. The French scientist's image of German science 1840–1919* (Gainesville, FL, 1972), 1–28, and Robert Fox, "The view over the Rhine: Perceptions of German science and technology in France, 1860–1914," in Yves Cohen and Klaus Manfrass, eds., *Frankreich und Deutschland. Technologie und industrielle Entwicklung im 19. und 20. Jahrhundert* (Munich, 1990), 14–24.

tain branches of electrical and mechanical engineering, even surpassed the achievements of Germany.

The explanation for France's weakness in the production of artificial dyestuffs, advanced by the Alsatian chemist Adolphe Wurtz in the wake of the World Exhibition (*Welt-Ausstellung*) in Vienna in 1873, was an early example of what was soon to become a common lament. In a review informed by a keen eye for the international perspective, Wurtz invoked the contrast between persistent French neglect of the facilities for science and Germany's heavy involvement in laboratories for teaching and research.[3] The relevance of this contrast was justified, for Wurtz, by the fact that, in dyestuffs, the gap between theory and industrial application was a narrow one. Hence, as he argued, success in manufacturing depended on there being sustained support for the academic "foyer scientifique" that in recent years had so profitably fed the dyestuffs industry in Germany: whereas in France, laboratories had remained much as they had been at the beginning of the century, provision in the German universities had improved out of all recognition. In a characteristically glowing comment on Germany, Wurtz wrote:[4]

> In every important center, laboratories have been founded, all of them magnificently built and liberally endowed. No resource is wanting, either for those who teach and cultivate science or for those who aspire to learn. The latter attend in large numbers....Those who have passed through and acquired a solid body of knowledge move on, year after year, into the ranks of society, some to follow careers as teachers, most of them to devote themselves to the industrial arts. In this way, the laboratories are at the same time schools of advanced science and nurseries for the fashioning of practical men....The expenditure that a country devotes to science and higher education is productive, therefore, and Germany has quickly reaped the fruits of its foresight.

In the forty years that followed Wurtz's warning, as France inexorably lost further ground in dyestuffs as she did in other "science-based" industries, such statements circulated freely within the French scientific community. In the hands of the chemist Charles Lauth, they helped to justify the agitation for a more industrially oriented curriculum that culminated in the foundation of the Ecole Municipale de Physique et de Chimie Industrielles in Paris in 1882,[5] and from the 1890s they were routinely in-

3. Charles-Adolphe Wurtz, *Progrès de l'industrie des matières colorantes artificielles* (Paris, 1876), 183–187.
4. Ibid., 184–185.
5. For Lauth's own account, see Charles Lauth, *Rapport général sur l'historique et le fonctionnement de l'Ecole Municipale de Physique et de Chimie Industrielles* (Paris, 1900).

voked in declarations of support for the new, technical institutes that were established in chemistry, electricity, and other applied studies in virtually all the science faculties of France.[6] In this latter movement, Albin Haller, professor of organic chemistry and, from 1890, director of the newly founded Institut Chimique in the Faculty of Science at Nancy, quickly became an influential voice. For Haller, the overriding cause of Germany's dominance in the chemical industry, especially in organic chemistry, was "l'organisation scientifique," which left German academic chemists with ample freedom to pursue the research that constituted the "haute culture" of their discipline, and with the material resources with which to do it.[7] No less importantly, it allowed students to exploit the German system of *Lehrnfreiheit* by individually fashioning their intellectual lives and so avoiding what Haller, evoking an earlier similar image of Hippolyte Taine's, portrayed as the French treadmill of mind-numbing cramming for examinations, in particular those attempted by aspirants to the Ecole Polytechnique.[8]

There was much in this and other parts of Haller's argument that resembled the positions taken by Wurtz and Lauth. But in certain respects, Haller's views, advanced in the wake of the triumphs of the German chemical industry at the World's Columbian Exposition in Chicago in 1893 and the Universal Exhibition in Paris in 1900 and subsequently as director of the Ecole Municipale de Physique et de Chimie Industrielles (1905–1925), went beyond the conventional statements about the benefits that would flow from the promotion of academic science. Written when they were, in the quarter of a century before the first world war, they reflected rapidly changing circumstances. For by now Haller could draw on the evidence not only of the escalating expansion of university-based chemistry in Germany in laboratories that he termed "factories of pure and applied science,"[9] but also of an integration of in-house research, adjacent to the sites of production, that had become the hall-mark of the German organic chemicals industry.[10] Victor Cambon, writing with characteristic fervor in the

6. On these institutes, see the studies cited in Chapter 3, refs. 72 and 73.
7. Haller, in *Ministère du Commerce, de l'Industrie, des Postes et des Télégraphes. Exposition Internationale de Chicago en 1893. Rapports publiés sous la direction de M. Camille Krantz. Comité 19. Produits chimiques et pharmaceutiques, matériel de la peinture, parfumerie, savonnerie* (Paris, 1894), 28, and in *Ministère du Commerce, de l'Industrie, des Postes et des Télégraphes. Exposition Universelle Internationale de 1900 à Paris. Rapports du jury international. Classe 87 – Arts chimiques et pharmacie. Rapport de M.A. Haller* (2 vols., Paris, 1902), *1*, xxx and xxxii.
8. Haller, in *Exposition Universelle Internationale de 1900* (ibid.), *1*, lxxx.
9. Haller, in *Exposition Internationale de Chicago* (ref. 7), 14, and in *Exposition Universelle Internationale de 1900* (ref. 7), *1*, xlv.
10. Haller, in *Exposition Universelle Internationale de 1900* (ref. 7), xlvii.

heightened atmosphere of the first world war, might insist that work emerging from such laboratories conspicuously lacked genius or inspiration. Nevertheless the "corps des chimistes de recherches" remained, for him, one of the essential sources of Germany's success.[11]

The growth of company laboratories in this sector of German industry was bred of a number of special circumstances. One was the size of the firms, which from an early stage tended to be significantly larger and to handle a broader range of products than was the case in other countries, making them better able to shoulder the risk of an investment in research, including the engagement of large numbers of highly trained staff, whose profitability was uncertain (see Table 4.1).[12] Another was the passing of the new German patent law of 1877, which replaced the disorderly and inefficient assortment of regulations inherited from the pre-unification states with a coherent structure that gave patent holders effective protection.[13] Thereafter, the commercial importance of seeking new products and of establishing a strategy for the registration and maintenance of patents was quickly recognized by all the large chemical manufacturers, especially in dyestuffs and pharmaceutical products. Within a few years, the benefits of the law were regarded as beyond question. In a lecture at the Imperial Patent Office in Berlin in 1893, Otto Witt, the dyestuffs chemist and professor of technical chemistry at the Technische Hochschule of Berlin, expressed in flamboyant terms what had already become received wisdom. Associating the German patent system with the soaring figures for the production and export of artificial dyestuffs since the mid-1880s, he presented the "blessings" of patent protection as "incalculable:" no tariff protection or any kind of state support for industry could have yielded comparable benefits.[14]

11. Victor Cambon, *Notre avenir* (Paris, 1916), 127–128.
12. For an indication of the large, and growing, size of German dye companies in the later nineteenth century, see Ernst Homburg, "The emergence of research laboratories in the dyestuffs industry, 1870–1900," *The British journal for the history of science, 25* (1992), 91–111, Table 2, 105.
13. The impact of the law of 1877 is widely recognized in the literature on research in the German chemical industry. See, for example, L.F. Haber, *The chemical industry during the nineteenth century. A study of the economic aspects of applied chemistry in Europe and North America* (Oxford, 1958), 202–203; John Joseph Beer, *The emergence of the German dye industry* [Illinois Studies in the Social Sciences, vol. 44] (Urbana, IL, 1959), 105–108; Georg Meyer-Thurow, "The industrialization of invention: A case study from the German chemical industry," *Isis, 73* (1982), 363–381; Homburg, "The emergence of research laboratories" (ref. 12), 104; Anthony S. Travis, *The rainbow makers. The origins of the synthetic dyestuffs industry in Western Europe* (Bethlehem, PA, London, and Toronto, 1993), 215 and 219.
14. Otto N. Witt, *Die deutsche chemische Industrie in ihren Beziehungen zum Patentwesen ...Acht Vorträge gehalten im Kaiserlichen Patentamt in Berlin* (Berlin, 1893), 141–142.

Table 4.1

Capital and number of employees in the artificial dyestuffs industry in France and Germany in 1900

	Share capital (in £ sterling)	Chemists	Engineers	Total no. of employees
France				
Société Anonyme des Matières Colorantes et Produits Chimiques de Saint-Denis (Poirrier and Dalsace)	360,000	25 (chemists and engineers)		440
Société Chimique des Usines du Rhône, Saint-Fons, Lyon (Monnet)	240,000	27	3	730
Germany				
Badische Anilin, Ludwigshafen	1,050,000	148	75	6207
Meister, Lucius & Brüning, Hoechst	850,000	129	36	3931
Bayer, Elberfeld and Leverkusen	600,000	145	27	4200
AGFA, Berlin	350,000	55	10	2026
Leopold Cassella, Frankfurt	?	80 (chemists and engineers)		2050

Note: Apart from the two French firms indicated here, the only other significant manufacturers of artificial dyestuffs in France were sub-sidiaries of German companies. There were several other German producers of artificial dyestuffs, in addition to those cited in the Table.

Source: Exposition Universelle de Paris en 1900. Les produits chimiques et les matières colorantes. Le blanchiment, la teinture et l'impres-sion des fibres textiles, special issue of *Revue générale des matières colorantes et des industries qui s'y rattachent*, 5e année, 1 Jan 1901, 555–592.

The history of research at the Bayer company, which accelerated conspicuously once the law of 1877 was in place, indicates the general growth in the activity that continued to the first world war. In 1879–1880, after some tentative first steps in the appointment of academically trained men in the mid-1870s, Bayer embarked on the engagement, in the laboratory of its Elberfeld works, of chemists whose explicit function was the search for new colors. In 1884, there followed a decisive appointment, that of Carl Duisburg as a relatively junior chemist. Still in his early twenties but armed with a doctorate from Jena and a year's experience of independent research (undertaken on behalf of the Bayer company) in Rudolf Fittig's chemical institute at the University of Strasbourg, Duisburg developed the red dye benzo-purpurine and a number of other new azo dyes that helped to rescue the company from its potentially ruinous dependence on what was by the 1880s, because of falling prices, the declining asset of alizarin.[15] With the value of such work amply demonstrated and the company's looming financial crisis of the mid-1880s averted, Duisberg's staff grew and soon moved, in 1891, into a purpose-built research laboratory on three floors and with individual work-spaces for two dozen chemists, costing £75,000.[16]

Isolated instances of what Ernst Homburg has described as "embryonic" research laboratories can be identified in Britain and France as early as the mid-1860s, when the Fuchsine company inaugurated a "laboratoire des recherches chimiques" under Charles Girard, a former pupil and assistant of Pelouze, in Paris.[17] But the phenomenon was pre-eminently German. According to Homburg and more recent studies by Carsten Reinhardt,[18] it

15. Haber, *The chemical industry during the nineteenth century* (ref. 13), 134–135 and 187–188; Beer, *The emergence of the German dye industry* (ref. 13), 72–93; and Meyer-Thurow, "The industrialization of invention" (ref. 13), 369–375.
16. For contemporary accounts of the laboratory, see "Das wissenschaftlich-chemische Laboratorium der Farbenfabriken vorm. Friedrich Bayer & Co. zu Elberfeld," *Chemiker-Zeitung, 17* (1893), 191–197, and Henry E. Armstrong, "The appreciation of science by German manufacturers," *Nature, 48* (1893), 29–34. Useful accounts of Duisberg's career at about this time are Carl Duisberg, *Meine Erinnerungen* (Leipzig, 1933), 28–73, and Hans-Joachim Flechtner, *Carl Duisberg. Vom Chemiker zum Wirtschaftsführer* (Dusseldorf, 1959), 49–132.
17. Homburg, "The emergence of research laboratories" (ref. 12), 96–97.
18. Carsten Reinhardt, "An instrument of corporate strategy: The central research laboratory at BASF 1868–1890," in Ernst Homburg, Anthony S. Travis, and Harm G. Schröter, eds., *The chemical industry in Europe, 1850-1914. Industrial growth, pollution, and professionalization* (Dordrecht, Boston, and London, 1998), 239-259. A fuller study of dyestuffs research at BASF and Hoechst is the same author's *Forschung in der chemischen Industrie. Die Entwicklung synthetischer Farbstoffe bei BASF und Hoechst, 1863 bis 1914* [Freiberger Forschungshefte, Reihe D: Wirtschaftswissenschaften, Geschichte, Bd. 202] (Freiberg, 1997), in which the review of recent literature on 7-17 is esp. relevant to the point being made here.

had its true roots in 1868, when Heinrich Caro, a former pupil of the Gewerbeinstitut in Berlin who had spent some years in the British dyestuffs industry (mainly as a colorist and chemist with Roberts, Dale & Co. in Manchester, working on aniline dyes), was appointed *mitleitender technischer Direktor* (essentially deputy technical director) by the recently formed BASF in Ludwigshafen.[19] Caro's career is a telling indicator of the process that was to take place in virtually all the leading German companies. By the mid-1870s, the balance of his activities, originally defined as laboratory work that was to be performed "from both a theoretical and a practical point of view," was already shifting from production to research, and by 1878 he had been appointed to the new post of "research director." Thereafter, the structure for research within the company developed steadily into a network of specialized laboratories, each focussing on work relevant to a specific range of products. At the heart of the system lay the base for Caro's own work, the *Hauptlaboratorium*. Rehoused in purpose-built premises in 1889 and subsequently enlarged to accommodate more than twenty chemists, this laboratory served a variety of interests that extended far beyond the routine control work of the production line to embrace the scientifically sophisticated tasks of managing new patents and searching for new products. Elsewhere, at Meister, Lucius & Brüning (Hoechst) the move to innovative research occurred somewhat less spectacularly. But the *analytisches Laboratorium* that was opened there in 1870 seems to have performed work that went beyond routine analysis. By 1880, work was in progress on the azo dyes, and Dr Christian Rudolph was leading a group pursuing the synthesis of indigo.[20] Thereafter, the restructuring of research in 1883 and the opening of a new central laboratory a decade later both signaled developments that were comparable with those at Bayer and BASF, though with the emphasis settling eventually on pharmaceutical products as well as on dyes.

The moderate pace at which research groups working on the examination and development of new products were introduced, first at BASF and Hoechst and then at Bayer, to complement the activities of existing laboratories devoted to production and quality control is striking. Despite the

19. Homburg, "The emergence of research laboratories" (ref. 12), 98–100; Reinhardt, "An instrument of corporate strategy" (ref. 18), 241-245, and the same author's *Forschung in der chemischen Industrie*, 65-77. On this and other aspects of Caro's life, see his obituary by August Bernthsen, in *Berichte der Deutschen Chemischen Gesellschaft*, *45* (1912), 1987–2042, and Anthony S. Travis, "Heinrich Caro, chemist and calico printer, and the changeover from natural and artificial dyes," in Robert Fox and Augustí Nieto-Galan, eds., *Natural dyestuffs and industrial culture in Europe, 1750-1880* (Canton, MA., 1999), 285-312.
20. Homburg, "The emergence of research laboratories" (ref. 12), 100–102.

early start about 1870, substantial investment in such groups did not occur until the late 1880s and 1890s, by which time the German organic chemical industry was already recognized as a world leader well on its way to becoming one of the wonders of the early twentieth century. It was now, but only now, that research in pursuit of innovation was finally sanctioned as a normal element in the strategy of any large company engaged in the production of fine chemicals, and only now that well-founded sites for the activity became common. It is true, as Homburg has observed, that we should beware of allowing any one company to stand as an exemplar for the whole industry;[21] in this respect, the unusual quality and accessibility of the Bayer archives may have had the effect of drawing undue attention to that company, at the expense of others. But the lateness of the integration of research as a well-funded designated activity in the organization of Bayer does allow cautious generalization. In 1897, six years after its opening, the main research laboratory housed fifteen chemists (working chiefly on azo and aniline dyes, intermediates, and indigo);[22] fifteen years later, however, the number of chemists engaged in research had risen to 57, spread between six laboratories, each with a specific function focussed on a particular range of products, now extending to pharmaceuticals, photography, and artificial rubber.

What came all too easily to be seen, in the wake of this expansion, as a direct and very simple correlation between the development of in-house laboratories and economic success captured the imagination of contemporaries and soon became a staple element of arguments of the kind that Haller advanced in his reports on the exhibitions of 1893 and 1900 in Chicago and Paris.[23] The German dyestuffs manufacturers themselves also contributed to the emerging consensus about the causes of their success; again international exhibitions, in particular that of 1900 (by which time Germany's output of artificial dyes had a value five times that of the whole of the rest of the world put together[24]), provided an opportunity for parading the novelties for which company chemists were responsible.[25] As a result, statements about the importance of training chemists in research and

21. Ibid., 91–92.
22. Meyer-Thurow, "The industrialization of invention" (ref. 13), 370–379.
23. See above, pp. 253-254.
24. Franz C. Huber, *Deutschland als Industriestaat* (Stuttgart, 1901), 385.
25. See, for example, the guide to the display mounted by the German chemical industry at the 1900 exhibition: *Weltausstellung zu Paris 1900. Sammelausstellung der deutschen chemischen Industrie* (Berlin and Munich, 1900), 68–80, 87–92, and 116–127 (the entries for BASF, Leopold Cassella & Co., and Friedrich Bayer & Co.), and the French edition, *Exposition collective de l'industrie allemande des produits chimiques* (Berlin and Munich, 1900), 68–81, 88–94, and 118–129.

experimental technique and the belief that the provision of well-founded industrial laboratories held the key to profitability, no matter what the economic context, were consolidated as the *leit-motivs* of public debate that they remained until the first world war. In these years, in Britain as in France, the sense of having been ousted from a position of eminence in leading sectors of the chemical industry reinforced this interpretation and fostered a stream of admiring accounts of German laboratories and calls to emulation, with respect both to the provision for research within the chemical industry and to the training in research that was offered in education.[26]

Although diffuse declarations of faith in research set the dominant tone of contemporary analysis, they did little to convey the diverse realities of everyday practice across the whole range of the chemical industry, and still less in other industries. Even in the research-led world of Bayer and the other German organic chemical companies on the eve of the first world war, the majority of laboratory chemists continued to devote far less time to innovation than to the more traditional humdrum round of quality control and of investigations directed to protective patenting whose main purpose was to prevent rival companies from developing new products. In the older-established sectors, such as heavy chemicals (the value of which remained far higher than that of artificial dyestuffs[27]), metallurgy, and machine construction, this balance in the prevailing conception of research was still more pronounced. Here, even when the value of modern company laboratories was referred to, far less store was set by originality than by improvements and adjustments aimed at efficiency and economy and by the efficient use of what were essentially well-tried procedures.

The tone of the literature in which this conception was advanced is illustrated by Victor Cambon's perceptive *L'Allemagne au travail* of 1909.[28] The author was a graduate of the Ecole Centrale des Arts et Manufactures

26. An early, but otherwise typical British example is the presidential address that W.H. Perkin delivered to the Society of Industrial Chemistry in London on 7 Jul 1885; see "The President's address," *The journal of the Society of Chemical Industry, 4* (1885), 427–438, esp. 438. For a similar sentiment, emphasizing the need to emulate German educational practices and "to create an atmosphere of research in our science colleges," see Henry E. Armstrong's opening address to the Chemical Section of the British Association for the Advancement of Science in Aberdeen in 1885, in *Nature, 32* (1885), 449–453 and 467–472, esp. 450–451.
27. About the turn of the century, the value of the artificial dyestuffs produced in Germany was about a ninth of the value of the total output of the German chemical industry. See Huber, *Deutschland als Industriestaat* (ref. 24), 385.
28. Victor Cambon, *L'Allemagne au travail* (Paris, 1909). Cambon pursued his attempts to alert his compatriots to the threat of Germany's economic power in a work of even greater passion published just before the first world war, *Les derniers progrès de l'Allemagne* (Paris, 1913).

who had pursued an industrial career, mainly in the production of glues and other animal products, and become a persistent critic of the administrative routine that he saw as the main impediment to French industrial success. Like most foreign commentators, he expressed unflagging admiration for the readiness of German industrialists, in chemical manufacturing as in other sectors, to invest heavily in laboratories. But Cambon's distinctive gloss was that, for him, the efficacy of Germany's contribution lay in a clear recognition of the distance and the difficulties that lay between the initial conception of a process or product and its full-blown exploitation.[29] Part and parcel of this perception was the breadth and flexibility of the German notion of the laboratory. Although Cambon readily accepted the pursuit of novelty in the laboratory as an important function, he saw this as only part of the story. Speaking of chemistry, he quoted admiringly the boast of one German manufacturer who claimed that of the 145 chemists on his payroll, roughly half worked in production and quality control and that, of the others who were engaged in what was broadly termed "research," only a tenth produced anything new.[30] With respect to metallurgy and mechanical engineering, Cambon's insistence that the term "laboratory" should be understood to include a wide spectrum of installations, including some that would more conventionally have been classed as workshops, emerged even more explicitly:[31]

> I mean the word laboratory in the broadest sense: a workshop where the properties of ferrous metals in traction and torsion are investigated, one where test constructions are carried out, another where the qualities of equipment under construction are checked, either as an end in itself or for the information of purchasers...all these are laboratories.

Cambon's book brings out the great variety of ways in which work carried out in laboratories—"laboratoires de contrôle des matières premières et des produits fabriqués, laboratoires de recherches pour les perfectionnements et les nouveautés"[32]—actually impinged on industry before the first world war (as it continued to do afterwards). And the vision that Cambon conveyed was amply endorsed by other testimony. With a typical show of national pride, the official catalogue of the German display at the Exposition Universelle in Paris in 1900 cited as "the principal cause" of the Empire's leading position in the electrical industry "the profound scientific training of the German engineer:"[33] almost 2000 graduates

29. Cambon, *L'Allemagne au travail* (ref. 28), 42–43.
30. Ibid., 45.
31. Ibid., 43.
32. Ibid.
33. *International Exposition Paris 1900. Official catalogue. Exhibition of the German Empire* (n.p., [1900]), 172.

of the Technische Hochschulen were said to be employed in the ten leading electrical companies. But the accompanying breakdown of the appointments held by these graduates showed that the overwhelming majority were engaged in management, construction, and the planning and maintenance of power stations, railways, and other installations. Only five per cent of them worked in laboratories, whose purposes were said, explicitly and significantly, to embrace "testing machines and apparatus" and "testing and examining material," as well as "examining new inventions."[34]

The fact that industrialists were so anxious to qualify their admiration for laboratory-based innovation and to stress the diversified realities of what constituted research does not imply, however, a reticence with regard to rigorous scientific methods. Quite the contrary. Metallurgy is a particularly good example of the way in which, about the turn of the century, a traditional industry adopted more sophisticated experimental techniques, in particular the new procedures of metallography. These procedures, incorporating the use of the microscope to examine prepared metal surfaces, had originated between the 1850s and 1870s with the work of Henry C. Sorby, a wealthy Sheffield amateur geologist, the German metallurgist Hermann Wedding, and Adolf Martens, the future director of the Koenigliche mechanisch-technische Versuchsanstalt (later the Königliches Materialprüfungsamt) in Berlin.[35] By the 1890s, following Floris Osmond's use of them in his studies of the composition and structure of steel, they had been integrated with chemical analysis and mechanical testing as a valuable tool in the examination of metals and alloys at the various stages of their preparation and application. The significance of the slow but steady growth in the use of metallography in the metallurgical industry was not lost on contemporaries: as they saw, even in a profoundly traditional industry such as this, precise instrumentation and scientific understanding were inexorably replacing the traditional prerogatives of the highly skilled foundry foremen whose experience allowed them to determine, often with extraordinary accuracy (though an accuracy that was soon to be surpassed by the new methods), the properties of the materials with which they were working.

No one articulated more clearly what was afoot and the way ahead than the French metallurgist, champion of industrial science, and engineer in

34. Ibid.

35. On this and the other scientifically based advances in metallurgy during the later nineteenth century, see Léon Guillet, *Les industries métallurgiques à l'avant-guerre. Leur avenir* (Paris, 1917), esp. 530–650, and the same author's *L'évolution de la métallurgie* (Paris, 1928). On the Koenigliche mechanisch-technische Versuchsanstalt and Martens's role in it, see the work by Martens and Guth cited in Chapter 3, ref. 20.

the Corps des Mines, Henry Le Chatelier.[36] The core of Le Chatelier's case, like that of his younger collaborator and fellow-metallurgist, Léon Guillet, was a conviction that industry in France should abandon the rule-of-thumb traditions of the shop-floor, which he saw as a deadening and deeply ingrained impediment to progress, and unreservedly embrace the methods of science.[37] Emphatically, once again, the case did not rest on any strong belief in the value of research as a source of fundamental novelty. It was far more important, in Le Chatelier's eyes, that industry should become broadly imbued with what he variously termed the "esprit scientifique" or the "méthode scientifique." Only employees who had imbibed these "scientific" attitudes in their education and training would be fitted to accomplish the reform. They would do so by steering the delicate middle course between, on the one hand, the mathematical, encyclopedic "esprit théorique" that Le Chatelier and Guillet portrayed as prevailing in the most prestigious higher technical schools and the universities in France, and, on the other, the narrowly practical concerns of certain educational initiatives, in particular the technical institutes of the faculties of science.[38] The failing of the former approach lay in its detachment from the realities of manufacturing. The latter, Le Chatelier believed, had diverted attention from the all-important *principles* of industrial practice through a preoccupation with the description of individual technological procedures. Such description, like rapid visits to factories or the fashion of bringing miniature blast furnaces and other equipment into the classroom, was no substitute for the prolonged direct engagement with the day-to-day problems of a factory that Le Chatelier saw as an indispensable element in the formation of any engineer.[39]

36. On Le Chatelier's life and opinions, see François Le Chatelier, *Henry Le Chatelier. Un grand savant d'hier. Un précurseur. Sa vie. Son oeuvre. Son temps* (Paris, 1968); also his last major work, *De la méthode dans les sciences expérimentales* (Paris, 1936).
37. For typically forthright statements, see Henry Le Chatelier, "La science dans ses rapports avec le développement économique du pays," *Comptes rendus hebdomadaires des séances de l'Académie des Sciences, 162* (1916), 663–666, and his preface to the book by Léon Guillet, *L'enseignement technique supérieur à l'après-guerre* (Paris, 1918), 9–28. By the time this book appeared, Guillet taught at the Ecole Centrale des Arts et Manufactures and the Conservatoire National des Arts et Métiers. Earlier, he had instituted a company laboratory at the factory of the motorcar manufacturer Dion-Bouton whom he joined shortly after leaving the Ecole Centrale in 1897.
38. Henry Le Chatelier, preface to Guillet, *L'enseignement technique supérieur à l'après-guerre* (ref. 37), 14–27, and Guillet's own remarks, ibid., 127–195. On the institutes, see above, 130-132 and 140-141.
39. Le Chatelier's prescriptions for the training of engineers and other industrial employees at various levels were elaborated in many of his writings. For a succinct

Laboratories were essential to the definition of Le Chatelier's ideal middle course of a *science industrielle* that would avoid the artificial distinction between *science pure* and *science appliquée*. It was in the laboratories of schools and universities that students would learn the experimental method, as opposed to either the abstractions that were drummed into the heads of *polytechniciens* or the empiricism and mere gathering of information that undermined the intellectual standing of the technical institutes; and it was then through quite separate industrial laboratories that the method would be applied and so penetrate to the heart of practice. While he applauded the research conducted in metallurgical laboratories and the papers published in the *Revue de métallurgie*, which he edited from its foundation in 1904, Le Chatelier remarked that by 1905 such work had yielded no new product or significant new process.[40] Its value, however, was not in doubt. It lay not in fundamental innovations but rather in the accumulation of what were individually minor contributions to such areas as the crystal structure of metals (through metallography and other photographic techniques), the chemistry of alloys, and the effect of the presence of various elements on the physical properties of steel. These contributions could not justify, by purely economic criteria, the maintenance of a laboratory by any but the largest and most opulent company. Yet they constituted the steady flow of discrete but essential scientific information and advice that had somehow to be maintained.

Le Chatelier preferred the foundation of laboratories that would serve whole sectors of manufacturing, and it was to the encouragement of such laboratories that he gave the highest priority in his pleas for the proper funding of industrial science. Public bodies, private companies, and individual benefactors should be encouraged to finance research that would then be available to all firms, rather than a single industrial interest. And the research should be of a very particular kind, with the emphasis on measurement and on projects with a sure rather than an open-ended outcome. To do otherwise, as he insisted in 1914, was profligate; the course of true discovery was simply too uncertain:[41]

> It is a mistake to provide a subsidy for research that could, quite properly, yield no result; you cannot stimulate discoveries to order. The essential

---

statement, however, see his *De la méthode dans les sciences expérimentales* (ref. 36), esp. 271-290.

40. Henry Le Chatelier, "Revue de la métallurgie en 1904," *Bulletin de la Société d'Encouragement pour l'Industrie Nationale*, 104e année (1905), 212–225, on 213.

41. Henry Le Chatelier, "Les encouragements à la recherche scientifique," a paper published in the series "Mémoires & documents" of *Le musée social*, no. 8 (Jul 1914), 88².

> goal of all subsidized research must consist in precise measurement directed at phenomena that are already known. It is while conducting measurements of this kind, in fact, that observant minds make their finest discoveries. It is a beginner's illusion to suppose that we can usefully take as the objective of our work the search for some yet unknown phenomenon.

Le Chatelier consistently espoused the ideals of organized research with clear, limited objectives from the late 1890s, when he took up this particular cause, until his death in 1936. By 1901, he had achieved a notable success under the aegis of the Société d'Encouragement pour l'Industrie Nationale, which implemented a program of patronage for research (mainly on ceramics, tanning, and metallurgy) whose results would be available to any interested manufacturer.[42] But the scale was small and, despite occasional collaborative ventures by trade organizations (such as the national federation of coal-mining interests, which established a laboratory for the study of safety in mines at Liévin in 1909[43]), it left France little better provided with laboratories by the time of the first world war than she had been at the turn of the century. The acceptance by the Académie des Sciences, in 1916, of the need for a national laboratory for physics and mechanics, on the model of the Physikalisch-Technische Reichsanstalt in Berlin or the National Physical Laboratory at Teddington, reflected a new openness to Le Chatelier's conception of research as a fount of useful, publicly available experimental data.[44] But once the crisis of war had passed, enthusiasm for a national initiative was lost, and France continued to rely on the Laboratoire Central d'Electricité and other, rather disparate testing laboratories, such as those of the Ecole des Ponts et Chaussées and the Conservatoire National des Arts et Métiers, whose income depended mainly on commissioned work for industry and public bodies.[45]

42. For a review of the work done under the society's auspices since its first preliminary ventures in the encouragement of research in 1891, see Le Chatelier's presidential address of 1904, in *Bulletin de la Société d'Encouragement pour l'Industrie Nationale*, 104 année (1905), 12–22, on 18–19. Le Chatelier's work for reform within the context of the Société d'Encouragement pour l'Industrie Nationale from the mid-1880s is well analyzed in Michel Letté, "Henry Le Chatelier et les réformes de la Société d'Encouragement pour l'Industrie Nationale, 1884–1906: une intégration progressive de la science industrielle comme modèle et principe directeur de l'innovation," *Sciences et techniques en perspective*, 2nd ser., *1* (1997), 347–404.
43. Le Chatelier, "Les encouragements à la recherche scientifique" (ref. 41), 79–80.
44. "Rapport de la commission d'action extérieure de l'Académie sur les laboratoires nationaux de recherches scientifiques," *Comptes rendus hebdomadaires des séances de l'Académie des Sciences, 163* (1916), 581–588 and 683. The committee that reported to the Académie was chaired by Le Chatelier.
45. On the laboratories at the Ecole des Ponts et Chaussées and the Conservatoire National des Arts et Métiers, see above, 139-140.

Not surprisingly, the issue of originality was even more marginal in Italy. Such was the backwardness of Italian manufacturing before 1900 that serious competition with other countries was impossible; the main problem being not so much to match the strides made by German or American manufacturers as to construct the very foundations of industry in Italy. For this reason, the practice of importing foreign products, especially machinery, metal goods, and chemicals, was seen as a price that had to be paid if Italy was ever to create a significant industrial base. This does not mean that Italian manufacturers made no attempt to establish a presence in certain advanced sectors. But in these cases, emulation and the assimilation of established technologies from abroad appeared both as the core of an essential learning process and, in the circumstances, as the safest route to eventual technological advancement.

Even as staunch and eloquent an advocate of modernization as Giuseppe Colombo, whom we have already cited for his work as the professor of industrial mechanics at the Istituto Tecnico Superiore in Milan and, from 1897 to 1921, director and inspiration of the school, shared these perceptions.[46] Colombo, it must be said, was by no means hostile to novelty. His contribution to the creation of the Società Italiana Edison and to the opening, in the center of Milan, of the Centrale di Santa Radegonda (1883), one of the first commercial plants for the generation and distribution of electricity in Europe, is illustrative of the many ways in which he strove to advance industry in Lombardy between 1870 and 1914. As a partner in a consulting firm specializing in the design of industrial plants, he was also directly involved in the modernization of a number of textile mills and other factories in the north of Italy.[47] Moreover, it was with Colombo's encouragement that Giovanni Battista Pirelli and Alberto Riva, both former pupils of his at the Istituto Tecnico Superiore, established successful businesses in technologically advanced fields—respectively in the manufacture of electric cables and of hydraulic turbines.

Despite his commitment to technical improvement, however, Colombo's program for the renewal of Italian industry gave a far higher priority to the implantation and diffusion of proven technologies than it did to the devel-

46. On Colombo's views, see Carlo Lacaita, "Giuseppe Colombo e le origini dell'Italia industriale," in Carlo G. Lacaita, ed., *Giuseppe Colombo. Industria e politica nella storia d'Italia. Scritti scelti 1861–1916* (Bari, 1985), 5–86. On the background in contemporary debate about scientific and technical education, see Anna Guagnini, "Higher education and the engineering profession in Italy: The *Scuole* of Milan and Turin, 1859–1914," *Minerva, 26* (1988), 512–548.

47. In 1872, Colombo became a partner in a consulting firm, Cantoni, Colombo and Mackenzie, that designed electrical plants and acted as an Italian agent for foreign electrical manufacturers.

opment of new ones: in order to bring Italian industry to the level of the pacemakers, he favored the judicious use of existing technologies and was always ready to import textile machinery, steam engines, and machine tools from abroad, if these offered the most appropriate solution.[48] In keeping with the view of the first director of the Istituto Tecnico Superiore, Francesco Brioschi, who believed that education should not allow itself to proceed in advance of industry, Colombo argued that the primary objective of an engineering school was to provide a foundation in current practice that was sound yet not so specialized as to limit the flexibility of students who had to make their careers in a rapidly changing and still unformed economic environment. In this program, which took the spread of know-how and the eradication of bad workmanship as its highest objectives,[49] research had a low priority and little effort was made to encourage it, despite the recognition of the importance of well-founded laboratories for the training of students and the provision of testing and other services for local manufacturers.[50] In the early years of Colombo's campaign, this neglect of research caused no great consternation, and it was only after the turn of the century and, more particularly, on the eve of the first world war that the state of the facilities for original work began to be perceived as a weakness. Now, as happened in France, concern about the subordinate position of the nation's industry with respect to other European countries began to gather momentum, and anxiety about the continued dependence on foreign technology, fanned by a rising tide of nationalism, began to transform attitudes. The promotion of research, both in industry and in academic institutions, suddenly assumed the status of an important cause, becoming a main item on the agenda of the Comitato Nazionale Scientifico Tecnico per lo Sviluppo e l'Incremento dell'Industria Italiana. Founded in 1916 as a joint initiative of academics and industrialists, the Comitato devoted its first meeting, in October 1916, to a survey of the state of applied research in Italy.[51] Not surprisingly, the results of the survey exposed the scarcity of modern laboratories.

48. Giuseppe Colombo, "Le industrie mecchaniche italiane all'esposizione di Torino (1898)," *Nuova antologia*, 77 (1898), 365–402; reprinted in Colombo, *Scritti e discorsi scientifici* (2 vols., Milan, 1934), 2, 1089–1101.

49. Giuseppe Colombo, "Le gallerie delle macchine del lavoro e del materiale ferroviario all'Esposizione Nazionale di Milano," a text of 1882, republished in Colombo, *Scritti e discorsi scientifici* (ref. 48), 2, 1060–1087.

50. See above, 210, 225-228.

51. "Storia del Comitato," *Atti del Comitato Nazionale Scientifico Tecnico per lo Sviluppo e l'Incremento dell'Industria Italiana*, 2 (1917), 3–13, and "Per i laboratori scientifici: Convegno dei professori di chimica e fisica," ibid., 23–31. The Comitato was dissolved in 1928, by which time the Consiglio Nazionale delle Ricerche (CNR), a governmental body created in 1923, had become the main agency for the promotion of research in Italy.

Broadening the context of industrial science

It is all too easy, in retrospect, for the views of Le Chatelier and Colombo on research to appear damagingly myopic and a crucial source of weakness in the sectors of engineering with which they were chiefly concerned. But such a judgement would take little account of the economic circumstances of France and Italy, which made an innovation-led strategy an unnecessarily risky and inappropriate option. No less importantly, it would overlook the limited extent to which success in the fast lane—in Germany and the U.S.—depended on an early commitment to innovative research. The roots of the success of these countries lay in a far more complex mosaic of interrelated and constantly evolving factors—economic, political, demographic, and technological. In certain configurations, this mosaic made it possible to devise a company strategy that would allow research focussed on innovation to yield substantial, sometimes spectacular, profit. Even faced with these examples, however, contemporaries seldom saw a research-led strategy of this kind as self-evidently the best option, and its advocates had constantly to justify a commitment to it. Buying in new ideas, whether in the form of patents held by independent inventors or through the absorption of smaller companies that held interesting patents, remained everywhere a route to technological innovation that offered the attractions of convenience and safety.

This observation builds on an expanding literature that in recent years has profoundly adjusted the historiography of the relations between science, technology, and industrial performance. One branch of the literature has a long tradition going back over sixty years to the work of Abbott Payson Usher and S. Collum Gilfillan, who argued from very disparate perspectives (Usher was an historian of technology, Gilfillan a sociologist) that technological change was largely the result of small cumulative improvements.[52] This interpretation has been revived and elaborated by a number of economic historians, most notably Nathan Rosenberg, whose influential books and articles have warned repeatedly against the prevailing tendency for historians of technology to focus on dramatic turning points and leaps forward.[53] Another, highly relevant literature is that concerning the

52. Abbott Payson Usher, *A history of mechanical inventions* (New York, 1929); S. Colum Gilfillan, *Inventing the ship. A study of the inventions made in her history between floating log and rotorship. A self-contained but companion volume to the author's "Sociology of invention"* (Chicago, 1935), and *The sociology of invention. An essay in the social causes, ways and effects of technic invention, especially as demonstrated historically in the author's "Inventing the ship"* (Chicago, 1935).

53. See esp. Nathan Rosenberg, *Inside the black box. Technology and economics* (Cambridge, 1982).

practices of certain industrial laboratories. Here, Leonard Reich and George Wise, both writing about the origins and early development of the American electrical industry, with special reference to such giants as General Electric and Bell, have elaborated the pertinent point that the very term "research" is obfuscating.[54]

Breaking with the stereotype of the successful research-led American corporation, Reich has described a laboratory at General Electric at the beginning of this century that devoted itself almost entirely to quality control and support for the company's patenting strategy. Certainly, it was the declared wish of General Electric's third vice president of engineering and manufacturing, Edwin W. Rice, that the company should create "a laboratory to be devoted exclusively to original research,"[55] but at the time Rice was speaking, in 1902, the full realization of his ideal remained some years ahead. Initially, the laboratory justified its existence (in the face of a somewhat skeptical board of directors) not so much by significant advances as by the more conventional activities of standardizing, enhancing reliability, and examining external patents that might be purchased with a view to creating a kind of protective belt from which rivals would be virtually excluded.[56] And it was only from 1907, when the young MIT- and Leipzig-trained chemist William Coolidge briefly took over from Willis Whitney as director of the laboratory (during the latter's illness) and then emerged, under Whitney, as a powerful influence on the strategy of General Electric, that research geared to innovation became the dominant activity and one that gradually allayed the suspicions of its critics. The turning point on the path to acceptance came when the long and tortuous process leading to a new generation of improved carbon and metal filaments for the electric light bulb eventually bore fruit, most notably with the development of the ductile-tungsten filament in 1912.[57]

54. Leonard S. Reich, *The making of American industrial research. Science and business at GE and Bell, 1876–1926* (Cambridge, 1985), esp. chapters 3 and 4, and George Wise, *Willis R. Whitney, General Electric, and the origins of U.S. industrial research* (New York, 1985), chapters 5–7.

55. Rice's announcement of the decision to establish the laboratory appeared in the annual report for 1901. See Thomas P. Hughes, *Networks of power. The electrification of western society, 1880-1930* (Baltimore, MD., and London, 1983), 165; Reich, *The making of American industrial research* (ref. 54), 71; and Wise, *Willis R. Whitney* (ref. 54), 97.

56. Hughes, *Networks of power* (ref. 55), 164–171; Reich, *The making of industrial research* (ref. 54), 69–79; Wise, *Willis R. Whitney* (ref. 54), 97–129.

57. Hughes, *Networks of power* (ibid.), 168–171; Reich, *The making of industrial research* (ibid.), 79–94; Wise, *Willis R. Whitney* (ibid.), 132–138.

The achievement served as an endorsement of the vision that the engineer Charles Steinmetz had championed from within General Electric since before the turn of the century. But historiographically, the success has its dangers. Above all, it risks diverting attention from the lateness of the emergence of the type of research that engendered it. No less importantly, it could obscure the fact that the prodigiously high cost of such activity placed it far beyond the means of all but the very largest and best-established companies. The new, innovative turn that research at General Electric took in the years just before the first world war would certainly not have been possible but for the growth and security whose strongest roots lay in part, in earlier, more routine work of the kind that had been conducted in the company's standardizing laboratory since 1896 and that continued to be pursued in other General Electric laboratories with specific production-oriented functions.

Important though recent historiographical trends have been in questioning the leading role that tends too easily to be ascribed to laboratory-based innovative research as a universal mainspring of technological success, there are respects in which they have not gone far enough. For however profoundly they have forced us to rethink our perceptions of the work that was conducted in industrial and academic laboratories, they have not displaced laboratories from the central position they still hold in some analyses of successful companies in the later nineteenth and early twentieth centuries. Structured laboratory facilities, substantial expenditure on them, team-work (as opposed to the efforts of individual inventors), and a close, formalized interaction between the laboratory and the factory are all commonly perpetuated in institutional histories as the classic ingredients of the achieving firm. And they have assumed a pivotal place in the "integrative hierarchy" of managerial functions that Alfred Chandler identifies as essential to the modern industrial enterprise.[58] Our view, as we have insisted, is not that such ingredients could not breed success: clearly they could and did—on a scale that increased significantly after 1918. The point is simply that there were alternative paths to development, and that this was as true of the fast-lane countries as it was of those in the slow lane.

This last point is well illustrated by the case of the vast and consistently profitable AEG, which made its way on the basis of only a modest and far from systematic commitment to fundamental innovation generated "in house." In this respect, AEG stood in marked and significant contrast with its main German rivals in the electrical sector, Siemens & Halske and the smaller company of Schuckert, both of which invested more substantially

58. Alfred D. Chandler, Jr., *Scale and scope. The dynamics of industrial capitalism* (Cambridge, Ma., and London, 1990), Introduction, esp. 31–34.

in independent research.[59] On the one hand, it was characteristic of Siemens & Halske that they sustained Werner Bolton for the six years, from 1896 to 1902, that he needed in order to arrive at his prototype of the tantalum filament lamp in the company's laboratory in Berlin;[60] it was characteristic, too, that just before the first world war the company decided on the construction of a new and much larger laboratory, the Forschungslaboratorium der Siemens & Halske AG und der Siemens-Schuckertwerke GmbH, which was eventually inaugurated in 1920.[61] On the other hand, it was equally characteristic, and no less commercially astute, that AEG preferred to pay Walther Nernst, then professor of electrochemistry at the University of Göttingen, £50,000 for the patent rights to his highly innovative lamp,[62] while concentrating the activities of its own laboratories on a conceptually less exciting program of incremental change aimed at enhancing reliability, reducing production costs, and adjusting products ever more precisely to the needs of customers.[63]

There is another, quite different reason why, in electrical technology, the emphasis on the commercial benefits that could be drawn from a purpose-built corporate laboratory remains misleading: it diverts attention from the innovation that could and did take place in the very different environment of the workshop, factory, or generating station. In such sites, the innovation was typically of an adaptive rather than a radically novel character. Nevertheless, work carried out in a context of this kind provided another channel by which practical know-how, often allied to sophisticated experimental procedures, could impinge, sometimes creatively, on indus-

59. On the laboratories that Siemens & Halske maintained from the earliest days of their existence in the mid-century, and on the development laboratories of Schuckert before the amalgamation with Siemens in 1902, see Georg Siemens, *History of the House of Siemens* (2 vols., Freiburg and Munich, 1957), 2, 25–36, and Ferdinand Trendelenburg, *Aus der Geschichte der Forschung im Hause Siemens* [Technikgeschichte in Einzeldarstellungen, no. 31] (Dusseldorf, 1975).

60. Trendelenburg, *Aus der Geschichte der Forschung im Hause Siemens* (ref. 59), 31–32, and Hughes, *Networks of power* (ref. 55), 166–168. It was another five years before the tantalum filament lamp became a major commercial success, and from 1914 the wolfram filament, developed in the U.S., was a powerful rival. Nevertheless, by 1914 over 50 million lamps with tantalum filaments had been sold. Bolton, who was still in his twenties at the time of his appointment, had studied at Leipzig under Wilhelm Ostwald.

61. Trendelenburg, *Aus der Geschichte der Forschung im Hause Siemens* (ibid.), 45–48.

62. Kurt Mendelssohn, *The world of Walther Nernst. The rise and fall of German science* (London and Basingstoke, 1973), 45.

63. The character of AEG's policy is brought out in Ulrich Wengenroth's biographical sketch of Emil Rathenau; see Wengenroth, "Emil Rathenau," in Wilhelm Treue and

FIG 23. The power station of the London Electric Supply Corporation at Deptford, 1889. From *The Engineer*, 67 (1889), 286. Reproduced with permission of the Institution of Electrical Engineers Archives.

trial practice. Once again, the point stands with respect to countries in both the fast lane and the slow lane, and it assumes special significance in accounts of the quarter of a century or so before the first world war, when the cutting edge of the electrical industry moved to the exploitation of high-voltage polyphase alternating current. This was a world of large generators, long-distance transmission lines, electric traction, and the rapid electrification of industry. In these areas, the new departures of the period 1890–1914 seldom originated in the laboratory, at least in the laboratory as it is traditionally conceived. They emerged far more commonly from the very different laboratory of the site, where new machinery was designed and produced and where experiments took the form of full-scale trials and test-runs.

A dramatic early example is that of Sebastian Ziani de Ferranti, whose efforts to establish a major supply station using the still untried technology of high-voltage alternating current as early as the 1880s has assumed the status of a legend.[64] In 1886, Ferranti installed a generating station at the Grosvenor Gallery in London in which two 700-horsepower alternators, designed and built by him, were harnessed to the production of alternating current at up to 2500 volts. A year later, carried along by this triumph (achieved when he was still in his early twenties), he moved on to an even bolder venture at Deptford on the banks of the Thames to the east of London. Here, with the support of a circle of enthusiastic backers in the London Electric Supply Corporation, he conceived a station of a completely new design in which, as in a laboratory, problems were generated and systematically addressed.

The grandiose scale of the Deptford project rested on an aggregate of daring technologies. The capacity was to be several times greater than anything that had been handled before. Both the size of the 10,000-hp alterna-

---

Wolfgang König, eds., *Berlinische Lebensbilder. Techniker* [Berlinische Lebensbilder, ed. Wolfgang Ribbe, vol. 6] ((Berlin, 1990), 193–209, esp. 203, 204, and 207.

64. On Ferranti's venture, see Hughes, *Networks of power* (ref. 55), 238–247. Other accounts include Gertrude Ziani de Ferranti and Richard Ince, *The life and letters of Sebastian Ziani de Ferranti* (London, 1934), 54–68; R.H. Parsons, *The early days of the power station industry* (Cambridge, 1939), 21–41; W.L. Randell, *S.Z. de Ferranti* (London, 1943); Percy Dunsheath, *A history of electrical engineering* (London, 1962), 159–168; Thomas P. Hughes, "The bold conception of S.Z. de Ferranti," *Consulting engineering, 3* (1963), 75–79; and Arthur Ridding, *S.Z. de Ferranti. A brief account of some aspects of his work* (London, 1964), 12–19; S.Z. de Ferranti, "Pioneer of electric power transmission: an account of some of the early work of Sebastian Ziani de Ferranti, D.Sc., F.R.S. (1864–1930)," *Notes and records of the Royal Society of London, 19* (1964), 33–41; Brian Bowers, *A history of electric light & power* (London, 1982), 144–150; and John F. Wilson, *Ferranti and the British electric industry, 1864–1930* (Manchester, 1988), 25–50.

tors and the length of the transmission lines (seven miles of underground cables) were unprecedented, and, with an output at 10,000 volts, the risk of breakdowns and accidents was a menacing presence. In a construction of such novelty, in which virtually every element was designed from scratch, there was no alternative to carrying out testing as work proceeded. The site, in fact, became a site of experiment, with all the precariousness that such a situation implied. A combination of untried theory and measuring and control equipment of palpable inadequacy at the voltages Ferranti was using forced on him the perilous expedient of trying his installation by running it at full capacity. Inevitably, the accidents came thick and fast. A fire, burned-out transformers, and power cuts all signaled failures that, in a normal laboratory, would have been simply the starting-point for further improvements in technique. But in the conditions that prevailed at Deptford, with the boundaries between experimental and normal operation blurred, the failures had grave financial consequences, quite apart from the pruning of Ferranti's plan that they entailed. By 1898, only ten years after the opening of the Deptford station, the London Electric Supply Corporation had lost £400,000, and it was not until 1905 that a first, rather paltry dividend was paid.[65]

The failure of Ferranti's grand experiment was due largely to an adventurousness that took him into areas of technology for which his background, based on on-the-job experience following his early employment in the experimental department of the Siemens Brothers works at Charlton in south-east London, was inadequate.[66] Success in the kind of venture he undertook had to await the maturing of a new generation of electrical experts whose activity was practical but also emphatically scientific, to the extent that it involved experiment, theoretical insights, mathematical techniques (of a kind specifically formulated by and for engineers), and the systematic analysis of production methods.[67] The new generation was composed predominantly of qualified engineers very different from the ap-

65. Hughes, *Networks of power* (ref. 55), 244–247, and Wilson, *Ferranti and the British electrical industry* (ref. 64), 48.

66. Ferranti received no formal higher education in engineering or science. After his secondary education at St Augustine's College in Ramsgate, he entered the Siemens Brothers works at Charlton in 1881. There he gained experience and a reputation for work on the design of dynamos before becoming chief engineer to the Grosvenor Gallery Company in 1886 and, in the following year, chief engineer and electrician to the newly formed London Electric Supply Corporation.

67. The attempts of contemporary scientist-engineers to investigate the problems encountered by Ferranti are examined (with special reference to John Ambrose Fleming's study of the Ferranti effect) in Sungook Hong, "Forging scientific engineering: John Ambrose Fleming and the Ferranti effect," *Isis, 86* (1995), 30–51.

plied physicists who had led the field of electricity in the pioneering days of the 1880s and were still active in many academic laboratories. Fashioned in the emerging courses in electrical technology that we described in Chapter 3, the engineers combined their ever more refined grasp of physical theory with experimental interests focussed firmly on the machines and installations themselves. The result was often work of real originality.[68] In this respect, the contributions of Gisbert Kapp (a graduate of the Polytechnikum in Zurich who worked as general manager of Col. R. E. Crompton's electrical works and laboratory in Chelmsford from 1882 to 1884, before becoming an independent consultant and, from 1905, the first professor of electrical engineering at the University of Birmingham[69]), General Electric's Charles Steinmetz (who used a brief stay at the ETH in Zurich in 1888–1889 to initiate himself in electrical engineering[70]), or AEG's Michael Dolivo-Dobrowolsky (one of the first students and then an assistant of Erasmus Kittler at the Technische Hochschule in Darmstadt in the early 1880s[71]) are exemplary.

Despite its importance for the electrical industry, the work of Kapp (on polyphase power and the associated transmission systems), Steinmetz (on alternating current motors and long-distance transmission), and Dolivo-Dobrowolsky (on the polyphase motor) grew in an evolutionary manner from existing practices, being creative without being fundamentally new. The contributions of most other electrical engineers of their generation shared this unspectacular character. Even when they went beyond the essentially routine areas of quality control, reliability, and the cost of production to embrace the innovations associated with the harnessing of electric power in new industrial contexts, they were usually adaptive rather than revolutionary and they drew on the know-how of mechanical engineering as they did on the theories of physics. This is not, of course, to denigrate their significance, still less their difficulty. Witness the example, described by Ulrich Wengenroth, of the exceedingly intricate processes that were involved in the economically important and technologically demanding task

68. On the development of distinctive electrical theories oriented to engineering rather than to physics, see Ronald R. Kline, "Science and engineering theory in the invention and development of the induction motor, 1880–1900," *Technology and culture, 28* (1987), 283–313, and the same author's *Steinmetz. Engineer and socialist* (Baltimore, MD, and London, 1992), 17–23.

69. David Gordon Tucker, *Gisbert Kapp, 1852–1922. First Professor of Electrical Engineering at the University of Birmingham. Appointed 1905; retired 1919* (Birmingham, 1973), 5–6, 12–20, and 25–27.

70. Kline, *Steinmetz* (ref. 68), 17–23.

71. Wolfgang König, "Michael Dolivo-Dobrowolsky," in Treue and König, *Berlinische Lebensbilder. Techniker* (ref. 63), 249–265.

of adjusting polyphase motors to the complex requirements of a textile mill.[72] As Wengenroth shows, the main challenge lay in the design of the intricate gearing and coupling mechanisms that were required in order to ensure a uniform speed irrespective of the number of machines in use. The solution called for a degree of mechanical ingenuity and flair and a skill in measurement techniques which, though far removed from the frontiers of physics, are too easily undervalued by comparison with the more theoretical and purely electrical side of the task.

So in seeking to identify the advances that took place in supply and distribution and in the application of electrical power about the turn of the century, we should not expect to find many instances of dramatically new departures. At least as much is to be learned from the case of the consulting engineer Charles Merz, who established and administered a fine unified system for the supply of electricity to Newcastle-upon-Tyne in the early years of this century on the basis of managerial flair, political agility, and a program of technological improvements focussed almost entirely on design and incremental changes in essentially well-tried practices.[73] In the light of Ferranti's experiences, Merz's policy might have been expected to persuade other British suppliers that innovations cautiously introduced could help to reconcile the ideal of technological progress with that of profitability. But, sadly for Britain, the guiding principle of the Merz model—that technology could only succeed if it and local conditions in legislation, politics, and the economy were finely adapted to one another and that all these diverse elements had to be adjusted to achieve a constructive coherence of purpose—bore little fruit outside the northeastern region where it had met such striking success. London, in particular, remained obdurate ground, despite the clear-eyed perception of its problems by Merz and his associate T. H. Minshall, the head of the London office of the consulting firm of Merz & McLellan. By 1914, technological advances in individual aspects of production and supply in the capital had failed to yield their full benefit (especially in the provision of power as opposed to lighting) because of the unfavorable context in which they were implemented.[74] The attempts that Merz & McLellan had made for a decade to persuade the London County Council to put in place a unified, centralized system of

72. Ulrich Wengenroth, "L'électrification dans l'industrie textile," in Monique Trédé, ed., *Electricité et électrification dans le monde* (Paris, 1992), 519–529, and "How they won the market: Electric motors in competition with steam engines, 1890–1925," in William Aspray, ed., *Technological competitiveness. Contemporary and historical perspectives on the electrical, electronics, and computer industries* (Piscataway, NJ, 1993), 270–286.
73. Hughes, *Networks of power* (ref. 55), 443–460.
74. Ibid., 249–261.

generation and distribution adapted to the needs of a large manufacturing city had fallen foul of politically motivated suspicion and managerial conservatism. As a result, London was left with the untidy and inefficient network of supply companies and plants that remained in place until well after the first world war.

Life in the slow lane

The countries of both the fast and the slow lane of electrical technology illustrate equally well that the relations between science and practice, or theory and implementation, cannot be understood without a clear appreciation of the diversity both of what constituted innovation and of the contexts in which innovation was bred and implemented. If we are right in the importance we attach to such tasks as trouble-shooting and the painstaking adaptation of electricity to customers' requirements and complex local circumstances, and to innovations in environments other than that of the conventional laboratory, the historiographical consequences are considerable. In our view, we should be ready to rethink the causes, arguably even the extent of the technological gap that opened between Germany and the U.S., on the one hand, and, on the other, Britain and other countries that proceeded at their various lesser paces in the slow lane. In such a reconsideration, which would take us to the heart of a number of long-running debates, notably about the supposed decline of British industry since the later nineteenth century,[75] we should need to review a number of favored interpretations.

The traditional emphasis on education and training is at the heart of one historiographical tradition that will certainly have to be qualified. In the last decade of the nineteenth century and in the early years of the new century, all the European countries that we have considered in this volume suffered from an excess rather than a lack of such graduates.[76] In these

75. A recent review of this particular debate, which was given new life by the books of Martin J. Wiener (*English culture and the decline of the industrial spirit* [Cambridge, 1981]) and Corelli Barnett (*The audit of war. The illusions and reality of Britain as a great nation* [London, 1986]), is in David Edgerton, *Science, technology and the British industrial "decline," 1870–1970* (Cambridge, 1996), esp. 15–28. For our own views on the issue of decline in British industry, see Robert Fox and Anna Guagnini, "Britain in perspective: The European context of industrial training and innovation, 1880–1914," *History and technology*, *2* (1985), 133–150, and "The flexible university: Some historical reflexions on the analysis of education and the modern British economy," *Social studies of science*, *16* (1986), 509–530.

76. For a comment on this problem as it affected France and Germany (in contrast with the more favorable circumstances in the U.S. and Russia), see Robert Fox, "France in perspective: Education, innovation, and performance in the French elec-

countries, the main problem did not lie in the number of young men who held degrees in technical subjects or industrially oriented science or in their reluctance to enter careers in manufacturing. Far more disturbing were the quality and the breadth of the knowledge that they possessed. A command of the theoretical foundations of the new industrial technologies, however sure, was no substitute for the know-how that could only be acquired at the cutting edge of practice.

In electricity, especially in the sector of high-voltage alternating current and the ever-expanding area of the industrial applications of electrical power, this kind of "engineering" knowledge was developed within industry itself—a respect in which electricity was no different from a number of other technologically advanced sectors. As Wolfgang König has pointed out, the knowledge passed from there into German higher technical schools through teachers who had gained working experience of the design and construction of machinery in the course of employment in electrical engineering companies. It is this process that justifies König's argument that, for this sector, the notion of a science-based industry should be replaced by its mirror image, that of an industry-based science.[77]

With regard to research, too—or more precisely to the emphasis on research as one of the defining characteristics of a successful modern industry—revisionism seems inescapable. In one interpretation, the integration of research in laboratories in industry before the first world war marked the extension, in a novel and visible form, of an established existing commitment to the quest for new industrially relevant knowledge. Even after the war, in fact, the laboratory remained the tip of a far more extensive iceberg of knowledge formation that remained rooted in the factory and the site. Now, as before the war, the larger companies in the fast-lane countries, where scale and diversification made experience and practical know-how more plentiful and more effective, were the main beneficiaries.

Notions of national mentalities also cry out for re-examination. In our analysis, not only in Britain but also in France and Italy, generally cautious attitudes towards independent innovation were not so much a mark of the congenital, debilitating unadventurousness that are often thought to have undermined industry in those countries. They were rather a calculated and often quite profitable response to the circumstances of weakness within

---

trical industry, 1880–1914," in Robert Fox and Anna Guagnini, eds., *Education, technology and industrial performance in Europe, 1850–1939* (Cambridge and Paris, 1993), 201–226, on 210–211.

77. Wolfgang König, *Technikwissenschaften. Die Entstehung der Elektrotechnik aus Industrie und Wissenschaft zwischen 1880 und 1914* (Chur, 1995), 227–296, and "Science-based industry or industry-based science? Electrical engineering in Germany before World War I," *Technology and culture*, *37* (1996), 70–101.

which strategies there had to be fashioned.[78] It is all too easy to forget that British, French, and (even more so) Italian industrialists found themselves obliged to take decisions, on research and innovation as on other matters, in economic, political, and demographic contexts very different from those that existed in Germany or the U.S.; even by the late 1890s, when the German, American, and Swiss lead in electricity was less marked than it was to become by 1914, they had a far more limited range of options and operated within constraints that reduced technological excellence to just one of a number of often conflicting considerations. In such circumstances, any concept of a universal ideal strategy in technology was hard to sustain.

The overtaking of France as a dominant force in the age of electricity was particularly brutal and hence, for our discussion of constraints and opportunities, instructive. Until the early 1880s, the country showed every sign of becoming a leader in the quest for innovation in electricity. Paris was the "city of light" well before Berlin became *Elektropolis*, and it provided creative figures of the caliber of Zénobe Gramme, Paul Jablochkoff, and Marcel Deprez with fertile ground for their inventive activity. By the 1890s, however, the pursuit of fundamental technological novelties had flagged visibly in France, and it was now the openness of the country to machinery and practices from abroad that was most striking.[79] In some contexts, such as the Compagnie Française pour l'Exploitation des Procédés Thomson-Houston (CFTH), the Société Westinghouse, and the Société Française d'Electricité AEG, the channel for the entry of foreign technology would be named subsidiaries of American or German companies. But more commonly, especially where German and Swiss technology was concerned, the passage into France, for both products and senior technical and administrative staff, came through companies trading under French names. Both Schneider at Le Creusot and the Société Alsacienne de Constructions Mécaniques at Belfort, for example, had agreements with Siemens & Halske that gave them the lucrative and safe right to exploit well-tried Siemens patents under varying degrees of German supervision. Likewise, the

78. For a discussion of France in this context, see Robert Fox, "Contingency or mentality? Technical innovation in France in the age of science-based industry," in Melvin Kranzberg, ed., *Technological education—Technological style* (San Francisco, 1986), 59–68.

79. The point is developed in Albert Broder, "La multinationalisation de l'industrie électrique française, 1880-1931: Causes et pratiques d'une dépendance," *Annales ESC*, 39e année (1984), 1020–1043, and Fox, "France in perspective" (ref. 76), 212–216. See also Peter Hertner, "Financial strategies and adaptation to foreign markets: The German electro-technical industry and its multinational activities, 1890s to 1939," in Alice Teichova, Maurice Lévy-Leboyer, and Helga Nussbaum, eds., *Multinational enterprise in historical perspective* (Cambridge and Paris, 1986), 145–159.

FIG 24. An advertisement for the Elektrizitäts-Aktiengesellschaft, formerly Schuckert & Co., showing the diversity of the company's interests, including its associated and subsidiary companies. From *International Exposition. Paris 1900. Official catalogue. Exhibition of the German Empire,* "Advertisements" (separately paginated), p. 73. Reproduced with permission of the London Library.

Compagnie Générale d'Electricité de Creil was effectively a subsidiary of Schuckert (from 1898 to 1902) and then, following Schuckert's collapse, part of the even more extensive international network of companies under the control of Siemens (see fig. 24).

In the years just before the first world war, when nationalist sentiment in France intensified, the extent of the technological dependence of companies with strong foreign links caused a good deal of alarm: in the case of the Compagnie Générale d'Electricité de Creil, for example, feelings ran so high that in March 1912 they prompted the resignation of four members of the Conseil d'administration who were held responsible for what critics described as "la germanisation à outrance" that had come to afflict the company.[80] But even in earlier years, when sensitivities were less keen, the dependence did not go wholly unchallenged. In 1898, the promoters of a proposed new company that was soon to emerge as the Compagnie Générale d'Electricité (CGE) presented their venture as a response to the humiliating invasion of French electrical manufacturing by foreign interests. At a time when there was only a handful of truly French companies exploiting French patents, the CGE declared that it would serve the national interest by providing an independent structure for the design and production of electrical equipment across the whole range.[81] In pursuit of its goal of turning the tide of technological subservience, it put in place an elaborate structure, involving a careful balance of specialized facilities, geared both to manufacturing and supply. Within three or four years, though, the bid had come to nothing. The initial dream of a proud technological autonomy had become a faded memory, and the CGE returned to the well-trodden French track of dependence. For all the initial flourish, there could be no concealing the fact that the company's products and services were now firmly based on patents leased from foreign firms, in particular the Swiss companies of Brown, Boveri and Alioth, and on the electrical machinery imported,

80. The growing resentment at the dominant position of senior German staff in French industry on the eve of the first world war is reflected in Louis Bruneau, *L'Allemagne en France. Enquêtes économiques* (Paris, 1914), esp. 219–242. On the Compagnie Générale d'Electricité de Creil (not to be confused with the CGE, which we discuss later in this paragraph), see the papers in the Archives Nationales (65 AQ G.207). The mounting tension at Creil and its consequences are described in Fox, "France in perspective" (ref. 76), 214 and 224. The "germanisation à outrance" (a phrase from an article in *La vie financière* for 8 Mar 1912) was seen as a consequence of the company's policy since 1902, when the previously happy association with Schuckert had been exchanged for a far less easy one with Siemens.

81. See the brochure announcing the formation of the company: *Compagnie Générale d'Electricité. Société anonyme en formation. Notice* (Paris, 1898), 3–4.

ready-made or partly completed, for erection in France.[82]

It is significant that the departure from the declared objectives of the CGE seems to have provoked neither opposition nor any particular sense of failure. In fact, the company's annual reports and the financial press suggest that the new policy strongly commended itself to shareholders, especially in the years of renewed recession between 1901 and 1903 that followed the boom of the late 1890s in which the CGE was born. As expert financial observers repeatedly reminded their readers, the CGE's decision to give a far higher priority to the installation of well-tested supply systems than to construction and the development of innovative machinery allowed it to survive successfully in a period when, as a young and decidedly vulnerable company, it would otherwise have been in serious difficulty.[83] The policy of providing modest but safe returns, advocated by the CGE's founder and senior administrator Pierre Azaria, had been vindicated by an economic performance that led to steady expansion from 1898, when the capital of the newly founded company stood at £400,000, to 1914, when it had reached £1m, a large sum by French standards, even if it was dwarfed by the capital of AEG, which was over seven times greater (see Table 4.2).

The state of dependence into which the CGE so quickly slid remained typical of the technological strategies of electrical manufacturers in France until the first world war. It was a condition that left little scope or incentive for flirtations with untried new technologies, still less for laboratory-based research. Yet it in no way diminished the importance of the advanced scientific knowledge and practical know-how that were called for if the unspectacular on-site work of erection, installation, and distribution was to yield profits of the kind that the CGE offered its shareholders. The Compagnie Electro-mécanique (CEM) was a smaller company that played similarly and with equal success to its strengths. From its foundation in 1885, it acquired a reputation for the ingenuity and reliability of the mechanical engineering on which any installation, whether for generation,

82. The policy of the company emerges very clearly from the annual reports and the rich collection of cuttings from the financial press in the file on the CGE in the Archives Nationales in Paris (65 AQ. G.160). The case of the CGE is discussed in Broder, "La multinationalisation de l'industrie électrique française" (ref. 79), 1031–1032; Pierre Lanthier, "L'industrie de la construction électrique en France avant 1914," in François Caron and Fabienne Cardot, eds., *Histoire générale de l'électricité en France*. Vol. 1. *Espoirs et conquêtes 1881–1918* (Paris, 1991), 700–703 and 712–714; and Fox, "France in perspective" (ref. 76), 214–215.

83. For evidence of the approval of the CGE's prudent policies, see (in addition to the published annual reports cited in the previous note) the supplement to the financial newspaper *L'information*, 23 Dec 1907.

Table 4.2

Capital of French and other electrical manufacturers, 1895–1913
(all figures in pounds sterling)

	1895	1900	1905	1910	1913
Breguet	?[a]	160,000	160,000	160,000	160,000
Compagnie Electro-mécanique	40,000	20,000	60,000	80,000	200,000
Compagnie Française pour l'Exploitation des Procédés Thomson-Houston	200,000	1,600,000	1,600,000	2,400,000	2,400,000
Compagnie Générale d'Electricité	—	600,000	600,000	720,000	1,000,000
Compagnie Générale d'Electricité de Creil	—	200,000	300,000	152,000	?[a]
L'Eclairage Electrique	66,000	160,000	160,000	240,000	465,000
S.A. AEG	—	40,000	40,000	800,000	80,000(?)[a]
S. A. Westinghouse	—	800,000	1,000,000	800,000	560,000
Société Française des Electrodes	—	40,000	40,000	48,000	96,000
Non-French companies					
AEG	1,056,000	2,880,000	4,800,000	5,200,000	7,440,000
Brown, Boveri	—	500,00	640,000	1,120,000	1,280,000

[a] Precise figures not available

traction, or the supply of power to an industrial plant, depended. Expansion never overstepped the bounds of prudence. But in the early years of the century, the CEM responded to demand and the revival of the economy by developing its new Le Bourget works as a center where turbines of high quality and sophisticated design were constructed and adapted to the alternators imported ready-made from Brown, Boveri in Baden. It was on this complementarity of an established indigenous base of fine French mechanical engineering and a capacity to adapt to judiciously selected new departures in electricity originating abroad that the company built its history of steady growth and prosperity.[84] The French Thomson Houston Company (CFTH) provides yet another example of technological caution allied to entrepreneurial acumen. In this case, from the foundation of the company in 1893 the key lay in the establishment of a near monopoly in the installation of electric power, for both industrial and domestic use, in the country's largest cities and, even more crucially, in electric traction, in which virtually all municipalities of any consequence displayed a keen interest from the mid-1890s.[85]

It would be wrong to suppose that the routes that these companies took to growth and profitability were easy ones: failures and disappointing performances were by no means uncommon, even among the more efficient and established of their competitors. The well-known engineering and construction firm, the Maison Breguet, paid no dividend of any kind between 1901 and 1907, and the descendant of the three companies that Edison founded in Paris in 1882, the Compagnie Continentale Edison, endured a succession of technological and economic crises through the 1890s and beyond, despite reducing the range of its activities to little more than those of an electrical supply company.[86] Although success was possible, therefore,

84. The point emerges from the brief company history, *La belle histoire de la CEM* (Paris, 1950), and more fully from the annual reports and newspaper cuttings in the Archives Nationales (65 AQ. M.118). Despite its high reputation, also founded on the construction of marine turbines of the Parsons type, the company expanded cautiously. Founded in 1885 with a capital of £12,000 (chiefly to fit out power stations), it still had a capital of only £80,000 as late as 1911. Growth just before the first world war, following the acquisition of the French branch of the Société Alioth, brought the capital to £400,000 in 1914. The closeness of the bonds with Switzerland is reflected in the presence of three Swiss members on the nine-man council d'Administration in 1913. For a further comment on the success of the CEM, see Lanthier, "L'industrie de la construction électrique en France" (ref. 82), 717–720.

85. Fox, "France in perspective" (ref. 76), 215–216. The strategy of the CFTH and its benefits are well analyzed in Lanthier, "L'industrie de la construction électrique en France" (ref. 82), 693–695 and 703–726.

86. Robert Fox, "Thomas Edison's Parisian campaign: Incandescent lighting and the hidden face of technology transfer," *Annals of science*, 53 (1996), 157–193, on 189–192.

its achievement called for strategies appropriate to economic, demographic, and legislative circumstances very different from those of Germany or the U.S. It called also for a sensitivity to local as well as national conditions. This local sensitivity was the key to the remarkable French success story of "white coal" (*houille blanche*), electricity generated in the hydroelectric installations that helped to transform the economy of the Alpine region region adjacent to Grenoble, in particular in electrochemistry, in the two decades or so before the first world war. Here, it was the routine, on-site skills and creative ingenuity of the existing mechanical engineering concerns of the region, most notably the firm of Neyret-Brenier, that allowed the potential of the advanced German and Swiss alternators to be realized.[87] The quality and sophistication of the imported elements in the system were essential, of course, as was the vision of engineers coming from outside the Dauphiné, most notably Aristide Bergès, a graduate of the Ecole Centrale des Arts et Manufactures whose paper-making interests brought him to the valley of the Isère in 1867. But no less important was the complementary role of such innovations as were introduced locally, in particular in the construction of the turbines and the handling of the huge pressures created as water came to be used in falls of ever greater height (rarely above 200m in the 1860s but already at 500m by the early 1880s and eventually to top 1000m) and to be transported from the reservoirs down to the generating stations in pipes of ever greater length and girth.

Many of the circumstances that made a substantial investment in laboratory-based research virtually inconceivable, and (at least in the short term) unnecessary, in the French electrical industry also existed in Italy. Like France, Italy between the 1880s and 1914 was overwhelmingly a receiver of foreign technology. Here, however, electricity was one of several sectors in which the potential scope for an independent contribution, both scientific and technological, was greater. As in France, where by the eve of the first world war senior positions were often held by foreign engineers and administrators or by French employees closely linked to parent companies abroad, so too in Italy general strategic decisions in many of the leading firms tended to remain in foreign hands. At a time when both finance and initiative came predominantly from outside Italy, in particular from Ger-

87. Victor Sylvestre, *La houille blanche* (Grenoble, 1925), 21–72; Henri Cavaillès, *La houille blanche* 2nd edn. (Paris, 1929), 53–71; Henri Morsel, "L'hydroélectricité," in Caron and Cardot, *Histoire générale de l'électricité*. Vol. 1 (ref. 82), esp. 596–610. On electrochemistry, in particular the relations between science and practice in this area, see also the thesis of Michel Caron, "Les bases scientifiques de la naissance des industries électrochimiques" (doctoral thesis, Ecole des Hautes Etudes en Sciences Sociales, 1987).

man and Swiss banks, the situation could hardly have been different.[88]

On this score, the successful Società Italiana Edison, established in 1884 with the main initial purpose of supplying electricity to the elegant cafés and restaurants of the Galleria Vittorio Emanuele in the heart of Milan, was atypical in that, despite the presence of Edison's former associate John William Lieb as senior electrical engineer, technological decision-making remained primarily in Italian hands.[89] Yet even in this case the impediments to a policy that might, in principle, have led the firm to engage in at least some form of independent innovation asserted themselves. In his capacity as the company's technical director, Giuseppe Colombo felt the constraints particularly keenly. When, in 1884, he proposed that the company should establish an experimental workshop for the design and production of its own electrical machinery and light bulbs, the idea was rejected; the company's main financial backer, the Banca Generale, judged it too risky.[90] As a result, the activity of the Società Italiana Edison remained essentially limited to the construction and running of electrical power stations in which the equipment was either imported or manufactured in Italy under licence. Eventually, in 1890, a small subsidiary company was set up to begin an independent program of development and production on a modest scale. But by now any prospect that this might lead to a serious engagement in innovation had disappeared. Foreign technology was flowing into Italy more abundantly than ever. This meant that a commercial strategy incorporating some degree of technological autonomy was no longer just the hazardous though feasible dream it had appeared to be in 1884; it had become quite unrealistic.

In the low priority that the Società Italiana Edison and other firms in the electrical sector still gave to innovation and *a fortiori* to laboratory research on the eve of the first world war, Italy was not significantly different from France: the ties to industrial or financial interests abroad made both coun-

88. Peter Hertner, *Il capitale tedesco in Italia dall'Unità alla prima guerra mondiale. Banche miste e sviluppo economico italiano* (Bologna, 1984), and "Il capitale tedesco nell'industria elettrica italiana fino alla prima guerra mondiale," in Bruno Bezza, ed., *Energia e sviluppo. L'industria elettrica italiana e la Società Edison* (Turin, 1986), 213–256. See also Luciano Segreto, "Capitali, tecnologie et imprenditori svizzeri nell'industria elettrica italiana: Il caso della motor 1895–1923," in Bezza, *Energia e sviluppo*, 175–210.

89. Lieb's role in the construction of the Santa Radegonda power station in 1883 and in the early years of the "Società Generale Italiana di Elettricità, Sistema Edison" (to give the company its full name) is outlined in *Lo sviluppo della Società Edison e il progresso economico di Milano*. Vol. 4. *Nel cinquantenario della Società Edison 1884–1934* (4 vols., Milan, 1934), 140–162.

90. Claudio Pavese, "Le origini della Società Edison e il suo sviluppo fino alla costituzione del 'gruppo'," in Bezza, *Energia e sviluppo* (ref. 88), 65–167.

tries, almost perforce, absorbers of foreign technology.[91] The fact remains, however, that in the installation of electrical machinery, Italian electrical engineers, again like their peers in France, found themselves exposed to demands that called for adaptive ingenuity, often amounting to genuine originality. Although the originality was not fundamentally innovative, we cite it to stress again the successes that could be achieved in circumstances of technological dependence. In northern Italy, the successes were bred of a demography and geography peculiarly favorable to the growth of the hydroelectric installations that became, from the start, the region's distinctive specialty in electrical technology. Of course, France too could boast significant achievements in hydroelectricity, but the distance that separated the main French areas of generation, in the Alps and the Pyrenees, from the most populated manufacturing region around Paris imposed a limit to development until an effective high-voltage transmission network was in place after the first world war.[92] In Italy, by contrast, the sources of hydroelectricity in the Alps and the main body of industrial and domestic consumers in Milan and Turin were far closer to one another and hence more immediately exploitable.

This geographical advantage enhanced the potential for profit and the consequent incentive to invest in the pursuit of greater efficiency in production and transmission. It is essential to the Italian success story in hydroelectricity that the incentive, far from being ignored, stimulated a major technological response. When the Società Italiana Edison opened a new generating station at Paderno, near Milan, in 1899, its alternators came (as they would have done anywhere in Italy or France) from abroad, in this case from Brown, Boveri, while the transformers were supplied by Ganz in Budapest, a company that by now had built famously on its established excellence in mechanical engineering to become a leading manufacturer of electrical equipment.[93] These derivative elements did not prevent the plant

91. The similarities between Italy and France have their limits, however. Henri Morsel has noted the particular openness of France to Swiss, as opposed to German, technological influences and finance (although the German presence was a powerful one); see Henri Morsel, "Les producteurs de l'électricité: leur organisation et leurs politiques," in Caron and Cardot, *Histoire générale de l'électricité*. Vol. 1 (ref. 82), 767–773. See also Lanthier's comment on the rapidity with which American participation in the capital of the newly formed CFTH, initially 40 percent, was eclipsed by French capital between 1893 and 1900; Lanthier, "L'industrie de la construction électrique en France" (ref. 82), 693.
92. The point is well made by Henri Morsel in his section on "L'hydroélectricité" (ref. 87), 610–612.
93. Guido Semenza, "L'impianto di Paderno," *Atti della Associazione Elettrotecnica Italiana* (1897), *1*, 121, and a publication of the Società Italiana Edison: *Note sui nuovi*

from being recognized as a showpiece. Paderno, in fact, was a technologically daring installation. Operating at 13,500 volts and with extensive transmission lines and an output of almost 8000 horsepower, it was at the time when it was designed in 1895 one of the largest hydroelectric power stations in Europe.

Although the origins of the individual core components of the Paderno plant are clear, it is more difficult to apportion the credit for the overall design and the system of distribution. A recent study of the novelty of Paderno has placed a greater emphasis on the contributions of the technical experts of Brown, Boveri and the other companies that supplied the machinery than on those of the technical staff of the Italian supply companies concerned.[94] Yet both Italian and foreign engineers played their part. In fact, the characteristics of the new generation of hydroelectric installation seem to bear the marks of an interaction between those who defined the requirements of the Italian companies and the representatives of the foreign suppliers of machinery. The former were sufficiently informed to perceive the practicability of what they wanted to achieve (even though their targets extended far beyond the realms of normal current practice), while the latter brought the know-how without which the goals could not have been realized.[95]

The quality of the individual components that were usually of Italian origin—in particular, the turbines, cables, and the insulating mechanisms—reflected the strength of the particular sectors of engineering in which Italian manufacturers managed to achieve good standards. These sectors succeeded in Italy through the relentlessly astute exploitation of the opportunities offered by the growth of the electrical supply industry. The case of the Milanese firm of Riva & Monneret, which made the turbines for Paderno, illustrates the kind of innovation on which success would be built. Drawing on long experience in the production of turbines, Riva & Monneret began, from the late 1880s, to specialize in the design and construction of these crucial components of large hydroelectric stations. The hallmark of

---

impianti della Società Generale Italiana Edison di Elettricità, 1895–1898 (Milan, 1899). See also *Lo sviluppo della Società Edison* (ref. 89), 163–182.

94. Roberto Maiocchi, "La ricerca in campo elettrotecnico," in Giorgio Mori, ed., *Storia dell'industria elettrica in Italia*. Vol. 1. *Le origini, 1882–1914* (Rome and Bari, 1992), 155–199.

95. As foreign visitors often pointed out, Italian power stations offered a remarkable range of the most advanced technology in the field. For a typical comment, see W. H. Donner, A. H. Foyster, R. S. Portheim, and J. T. Morris, "Report of the Committee on Traction, Light, and Power Distribution on the visit of the Institution to Italy in 1903," *Journal of the Institution of Electrical Engineers*, 33 (1903–1904), 838–843.

their approach was a practice of designing each machine individually, so as to tailor it to the conditions of the plant in which it was to be used.[96] The work, which required a sensitivity to a variety of factors—such as the nature of the supply of water, the interface between the turbines and the generators, and the profile of the output that was required—was adaptive in nature, and it owed virtually nothing to laboratory research. Yet it was research in precisely the low-key manner whose importance for the development of electrical technology we are seeking to highlight. Riva & Monneret made it the foundation of their achievement and of their impressive record in competing with foreign manufacturers and securing orders for equipment within the range of their specialities, in particular for their turbines. The fact that they were called in to supply two 3000-hp turbines for the first plant of the huge Niagara Falls project, completed in 1899, is entirely consistent with their reputation and the acknowledged quality of their products.[97]

Equally important and striking were the achievements of another Italian company that contributed to the Paderno project: Pirelli. Founded in 1873 by Giovanni Battista Pirelli as the first Italian producer of rubber goods, the company subsequently diversified to embrace the manufacture of the telegraphic and electric cables for which it soon acquired an international reputation. But until after the first world war, Pirelli maintained only a small laboratory at its headquarters in Milan, devoted almost entirely to testing and product control. The incremental but, in the end, substantial technological advances that it made in the design of electric cables for the transmission of alternating current at increasingly high voltages rested therefore on experience gained in the field.[98] In this work and Pirelli's other pre-war success, the chief architects tended to be (then as now) relatively little-known figures, unsung heroes such as the director of the cable department, Leopoldo Emanueli, and the firm's technical director, Emanuele Jona. Both were highly educated graduates of Italian engineering schools

96. The firm was established in 1872 by Alberto Riva, a graduate of the Istituto Tecnico Superiore of Milan and a former pupil and protégé of Colombo. In 1887 Ugo Monneret de Villard, who had studied at the ETH of Zurich and worked on traditional hydraulic turbines in Switzerland, joined the firm as technical director. See Guido Ucelli, *La Riva in cento anni di lavoro, 1861–1961* (Milan, 1961); Carlo Sicola, *125 anni per l'energia* (Milan, 1988); and Giorgio Bigatti, "Commercianti e imprenditori nella Milano post-unitaria: Le origini della Riva, 1861–1896," *Storia e società, 11* (1988), 53–99.
97. Bigatti, "Commercianti e imprenditori" (ref. 96).
98. Alberto Pirelli, *La Pirelli. Vita di una azienda industriale* (Milan, 1946), and Hector Sacchetto, *History and development of the oil filled cable* (published privately, n.p., 1969), chapter 2.

(Jona, in addition, had studied at the Institut Montefiore in 1885), and hence were representatives of the new breed of formally trained electrical engineers whose impact from the 1890s has yet to be fully recognized.

Yet another success-story, even more remote from the world of laboratory research, was that of Ercole Marelli.[99] After receiving virtually no formal education, Marelli was apprenticed to one of the earliest electrical firms in Italy, the largely unsuccessful Tecnomasio of Milan, before establishing his own modest factory for the production of small electrical appliances in 1891. Among these appliances were electric fans of the type that was just beginning to be imported, for Italian domestic use, from the United States. As the range of Marelli's products grew, in a mere seven years, to embrace not only domestic fans but also large and technologically more ambitious electrical ventilating systems for industrial plants, the success of the firm became a source of intense pride. By 1911, the official company history looked back on an expansion that had taken the workforce from forty in 1897 to its present 1500 and on a vigor that had resulted in a flourishing export market and in a manufacturing plant equipped with the most modern machinery.[100] Yet the only source of technical know-how to which the history referred was the drawing-office and its senior staff of engineers working on problems of design related to the immediate needs of production. It was from this problem-solving approach that innovation emerged, to the visible economic benefit of the company.

In conclusion, there remains a point of general interpretation that it is important to elaborate. We have argued that the readiness or reluctance of companies to invest in research laboratories as sites of institutionalized innovation before 1914 cannot, on their own, provide a satisfactory explanation for economic success or failure. We should make it clear, however, that we do not underestimate the effect that the close interaction between science, engineering, and industrial practice had on the growth of the new so-called science-based industries that emerged in the late nineteenth century. The interaction was an essential characteristic of the period in most sectors of industry, and we recognize that we could have given added substance to this point by exploring case-studies of other contexts in which the interaction occurred: David Cahan's account of the development of the ultramicroscope at the Zeiss Werke in Jena and Hans-Liudger Dienel's study

99. On this case, see *Vent'anni di vita della Ditta Ercole Marelli & C., Milano. 1891–1911* (Milan, 1911), and Stefania Licini, "Ercole Marelli e Tecnomasio Italiano dalle origine agli anni Trenta: un tentativo di comparazione," *Annali di storia dell'impresa, 5/6* (1990), 299–321.
100. *Vent'anni di vita della Ditta Ercole Marelli* (ref. 99).

of Carl Linde's contribution to refrigeration technology would have been just two obvious choices from a burgeoning secondary literature.[101] Our aim, though, has not been comprehensiveness but rather the presentation of the general point that so long as historians give pride of place to the contribution made by research laboratories, their vision of the history of innovation and technological change will be a partial one.

The broadened perspective that we advocate raises particular historiographical problems. It calls for as great a degree of interdisciplinarity as we can muster, and it has obvious repercussions for the ways in which we interpret and explain the diverse levels of performance in different industries and different countries. Take, for example, one central question that emerges implicitly from our discussion of technological and business strategies in our slow-lane countries and from many other discussions of backwardness and stagnation. In spite of all we have said about the importance of forms of research conducted in the engineering workshop, on the production line, or on the site, might it yet be that a greater investment in laboratories and a serious engagement in fundamental innovative activity would have allowed Britain, France, or Italy to make good the ground in electrical technology that, in their different ways, they lost from the later 1880s? Our view of the period would suggest not. For activity of this kind to be effective, the context had to be one that made laboratory-based novelty technologically and, wherever possible, economically rewarding. Where such a context was absent, no amount of investment in facilities for research could have made up for the lack. Research on its own, in other words, would have offered no short cut to profitability, no matter how creative it might have been.

In short, the difference between success and failure cannot be accounted for, in our period or (for that matter) in any other, by monocausal explanations. Instead, we have to enter into the complex realities of multiple causes that assume their explanatory weight only they are considered in relation to one another and as part of a broader whole. In that mosaic, the elusive notions of adaptation and creativity in responding appropriately to the needs of the market, whether long-felt and clearly identified or stimulated more abruptly by a new product or application or by a lowering of price,

101. David Cahan, "The Zeiss Werke and the ultramicroscope: The creation of a scientific instrument in context," in Jed Z. Buchwald, ed., *Scientific credibility and technical standards in 19th and early 20th century Germany and Britain*, (Dordrecht, Boston, and London, 1996), 67–115. Hans-Liudger Dienel, *Ingenieure zwischen Hochschule und Industrie. Kältetechnik in Deutschland und Amerika* , 1870-1930 (Göttingen, 1995).

necessarily have a central position. As in Franz Huber's contemporary commentary on Germany's powerful position as an industrial nation in 1901,[102] commercial strategies emerge from such an analysis as being at least as important as science.

102. Huber, *Deutschland als Industriestaat* (ref. 24). As professor at the Technische Hochschule in Stuttgart and a noted authority on the social and economic problems of modern industrial society, Huber gave a conspicuously balanced view of Germany's rise to the status of a dominant industrial power in the later nineteenth century. His analysis brought out the favorable political and economic circumstances as well as the country's technological and scientific strengths.

CONCLUSION

Our aim in this book has been to examine the way in which research, and more specifically research with an orientation towards industry, came to occupy a prominent place on both academic and industrial agendas during the nineteenth and early twentieth centuries. In doing so, we have adopted a transnational perspective, focusing on European countries that differed significantly from one another both in their economic and social circumstances and in their institutional traditions. We fully acknowledge that these particular national backgrounds played an important role in fashioning, within the various countries, the character of the research that was performed and the pace of its integration in institutions of higher scientific and technical education. By the same token, we also recognize that national backgrounds exerted a powerful influence on the processes that led to the adoption of research practices in industry. We see these points as broadly squaring with prevailing current interpretations, which tend to bring out the diversity of the outcomes of the processes we describe. But we do harbor significant reservations about the implications that are commonly drawn from these interpretations. Here, our reservations reinforce the main conclusions we reached in earlier studies of the history of higher technical education,[1] and it is to this common core of interpretation, in particular as it applies to the electrical industry, that we devote the following final comments.

In all the countries we have treated, both those in the fast lane that took the lead in electricity and those in the slow lane that lagged behind, a new emphasis on the importance of laboratory equipment and laboratory-based pedagogy and on research as a function of higher education (embracing both the work of professors and the task of training pupils in research) came to the fore in the later nineteenth century. As we showed in Chapters

1. See Robert Fox and Anna Guagnini, "Britain in perspective: the European context of industrial training and innovation, 1880–1914", *History and technology*, 2 (1985), 133–50, and "Introduction", in Robert Fox and Anna Guagnini, eds., *Education, technology, and industrial performance in Europe, 1850–1939* (Cambridge and Paris, 1993), 1–9.

2 and 3, physics departments and schools and departments of engineering alike tried, for different reasons and with different degrees of success, though always with alacrity, to seize on the new opportunities that were offered by the sudden and spectacular growth of electrical manufacturing and utilities. Everywhere, the move provoked reactions ranging from coolness to strong objection in the more traditional sections of the academic community, with resistance being most marked when university-based science departments were involved. Nevertheless, the will to establish bonds with the world of production—a will that, especially in the case of physics, was often bred as much of expediency as it was of profound conviction—almost invariably won out, and, once established, such bonds were resolutely exploited as central elements in the strategies of universities and higher technical schools.

Wherever academic laboratories were created or reorganized in such a way as to encompass the advancing world of electrical technology, the material facilities and the capacities of the staff were proudly advertised and placed at the disposal of industrial concerns. Within the limits of what different institutions could offer, such laboratories were used for a range of testing and other functions, and professors acted as consultants; in this way, through the income and interest that were generated, it became possible to improve equipment and to extend the space available for research. Often academic staff would also conduct experiments within industrial plants, while machinery from plants would not uncommonly be installed and used for testing and study in laboratories. By 1900, this blurring of the boundaries between the domains of higher education and manufacturing had come to represent a trend in all the industrialized countries of Europe, and thereafter it was taken further by the increasing tendency, within the academic world, not only for engineers but also (albeit to a lesser extent) for physicists to engage in patenting in their own right.

Although we have insisted on the general pattern of openness towards industry on the part of the academic community, we have also pointed to the microstructure that existed behind this pattern: as we have argued, the chairs of electrical engineering and the laboratories in which professors and their assistants in their disciplines pursued research always bore the mark of the particular institutional background from which they emerged and the resources they attracted. This point about microstructure is important, since it bears crucially on the central question of the extent to which the resulting unevenness in the body of knowledge and expertise that universities and higher technical schools could offer affected the industrial performance of the countries in which they were situated. In *Laboratories, workshops, and sites,* we have tried to address the question by considering industry as something other than a mere recipient of the results achieved within academic institutions and a beneficiary of the various traditions of research that such institutions engendered. Instead of a unidirectional model of research feeding industrial practice, we have taken a broader view of both the nature and the sources of the improvements that characterized the electrical industry, especially in the years of its most dramatic expansion between the 1890s and the first world war.

We ascribe a central role in these years to low-key research of the kind we discuss in Chapter 4 (with reference to parts of the chemical as well as the electrical industry). It is an essential part of our argument, in fact, that in electricity research in this sense of an intensely practical problem-solving activity was pursued, to a degree that has not been fully recognized, both in countries that were at the technological cutting edge of the "electrical revolution" and in those that were not. We admit, of course, that in the crucial task of adapting technologies to the requirements of clients and to the opportunities that presented themselves at different times and in different places, large electrical engineering firms with international interests made a substantial contribution through their own work in design and installation. But, even in the case of the firms in this category, a significant part of the task of adjustment was undertaken locally, using expertise acquired through first-hand knowledge of the context in which the new machinery and techniques were to be used. As Nathan Rosenberg has argued, such adjustment to local circumstances was almost invariably difficult, and it can certainly not be regarded as just a superior form of copying.[2] It called for competence of a high order in engineering and problem-solving, and, when successfully conducted, it yielded substantial economic benefit, even in unpromising circumstances. In Chapter 4, for example, we saw how a capacity for finely gauged adaptation provided the springboard for the by no means negligible success of the companies in the slow-lane countries that managed to exploit unfilled niches in the electrical market from the end of the century.

In drawing attention to forms of research aimed primarily at adaptation and to tasks of improvement that had little to do with more fundamental kinds of innovation, we have highlighted a realm of activity common to both the fast and the slow lanes of industry. Yet it is precisely the importance of innovation in the sense of small incremental developments, allied to the wealth of technical know-how and gadgetry that was required in order to sustain it and make the best use of it, that helps to explain the accelerating divergence between the fast lane and the slow lane of industry, in particular after 1900. Here, the circumstances in which such research

2. Nathan Rosenberg, *Perspectives on technology* (Cambridge, 1973), 141–210 (Chapter 3 on "Diffusion and adaptation"), and *Inside the black box. Technology and economics* (Cambridge, 1982), 245–91 (Part IV: "Technology transfer and leadership: the international context").

was pursued were decisive. For although it was pursued throughout the industrialized world, the levels at which it was conducted were determined by the nature of the problems to be addressed and by the means (both physical and conceptual) that were available in order to tackle them. Inevitably, the more advanced the technological objectives in question were, the more advanced was the level of the work that was carried out. The technologically most fruitful context, in other words, was one in which a wide range of experience, acquired in day-to-day practice, was available, and in which, no less importantly, the proximity of the frontiers of technology created the stimulus for problems of a particularly challenging nature to be tackled.

It could be argued, therefore, that one reason why the gap between the fast lane and slow lane in electrical engineering widened from the 1890s was precisely because the laboratory that really mattered was so close to the production site. In 1890 the gap between the electrical engineering industries of different parts of Europe had still been relatively small. But those countries that had secured even a modest lead by then were in a position to forge ahead ever more rapidly over the next two decades. The result was a pattern of rapidly widening gaps between the countries that we have examined. The emergence, after 1900, of the first research laboratories that were devoted not only to testing and quality control but also to innovation further contributed to the process, and it is in no sense the purpose of *Laboratories, workshops, and sites* to play down this contribution. But the function of the new generation of laboratories was one of consolidation and the enlargement of a lead secured initially by other means. It was not its root cause.

Finally, a last comment on methodology and implications. We cannot stress too strongly that the approach we have adopted, based on a multifaceted and essentially interdisciplinary analysis, only assumes real significance in a comparative setting. At least in principle, historians have acknowledged this painfully obvious point, and projects and multinational ventures that reflect its acceptance are beginning to bear fruit. But we have yet to convey the complexity of the past to the makers of policy who are so ready to draw historical lessons in isolation from the context that alone gives them meaning. The history of applied research and technological innovation is an area that has been flagrantly misused in this way, and it has been at least a subsidiary purpose of our book to point to the dangers of a selective reading of the past colored by the preoccupations and preconceptions of the present and by an indifference to the need for the precise contextualization of historical evidence. We are concerned, for example, that our contention that the efficacy of laboratory-based research before the first world war was uncertain should not be read as necessarily implying doubt about its efficacy in the very different circumstances of our own day: indeed, our argument accommodates without difficulty the fact that, a century ago, a research-oriented industrial policy, pursued in the right context, was capable of success and that *a fortiori* such a policy may succeed today. The conclusion we wish to stress, however, is that such a research-oriented policy has never been a panacea to be adopted in all places and at all times. At least that degree of uncertainty has to be acknowledged, along with our point about complexity, if historians are to debate the significance of history not only among themselves but also with contemporaries in search of easy solutions to the difficulties of the often flagging economies of the end of the twentieth century.

INDEX